Ethical Transformations for a Sustainable Future

Ethical Transformations for a Sustainable Future

Peace & Policy
Volume 14

Olivier Urbain and Deva Temple, editors

Routledge
Taylor & Francis Group
LONDON AND NEW YORK

First published 2011 by Transaction Publishers

Published 2017 by Routledge
2 Park Square, Milton Park, Abingdon, Oxon OX14 4RN
711 Third Avenue, New York, NY 10017, USA

Routledge is an imprint of the Taylor & Francis Group, an informa business

Library of Congress Catalog Number: 2010004007

Library of Congress Cataloging-in-Publication Data

Ethical transformations for a sustainable future / Olivier Urbain and Deva Temple, editors.
p. cm. -- (Peace and policy ; v. 14)
Includes bibliographical references.
ISBN 978-1-4128-1445-4 (alk. paper)
1. Environmental ethics. 2. Environmental protection--International cooperation. 3. Environmental protection--Moral and ethical aspects. 4. Sustainable development. I. Urbain, Olivier, 1961- II. Temple, Deva.

GE42.E798 2010
179'.1--dc22

2010004007

ISBN 13: 978-1-4128-1445-4 (pbk)

Contents

Rethinking Education and Technology

Preface

Olivier Urbain

Our planet does have a future, but it may not be one suitable for the human species. How did we get to this point? After learning how to use water and produce food surpluses, we invented agriculture, created civilizations, designed machines to help us, and learned how to communicate at the speed of light. Where did it all go wrong? Human life on Earth is no longer sustainable. Climate change is catching up with us, the financial and economic crisis is threatening jobs everywhere, we still have 25,000 nuclear weapons of which 3,500 placed on hair-trigger alert, waiting to explode on a minute's notice. Natural resources are rapidly dwindling, fresh water is becoming increasingly scarce, peak oil is drawing near.... We use our grain to feed livestock so that a minority of us can eat meat while the majority is struggling with food shortages and about one billion, almost one sixth of the world's population, starve. The gap between the rich and the poor is widening, the Millennium Development Goals are becoming a distant dream, massacres of civilians go unpunished in Africa and the Middle East, in South Asia and elsewhere ... what can we do?

What should be done has become obvious for many of us, ordinary people and heads of states alike: we need to move towards a world free of nuclear weapons, use renewable resources and green the economy, reform the UN and give ourselves the effective and humane global governance we deserve, develop conflict transformation skills to avoid violence, design new financial, economic, educational, legal, scientific, political systems that will ensure a different way of living on this planet, a sustainable way.... So, where do we start?

The contributors to this volume believe that transformations at the ethical level are necessary in order to start moving towards a future that will be sustainable for life on Earth. The most fundamental ethical shift is to be able to see that all life is interconnected and that our political, economic, and social systems must be based on sustainability—for all living and non-living systems, not merely for human convenience. While not spectacular or newsworthy, an inner shift within each of us is the most solid point of departure towards sustainability. The nine chapters in this volume give concrete examples of how this can be accomplished. We hope you find them intellectually challenging and philosophically inspiring.

Olivier Urbain
Director, Toda Institute
Tokyo, Japan
June 16, 2009

Introduction

Olivier Urbain

It has become increasingly apparent that humanity's complex social, political, and economic systems are incompatible with the finite capacity of the Earth to replenish resources and absorb wastes. If the planet is to continue to be habitable for us, our descendants, and for the myriad web of other creatures that make life possible, we must transform human systems so that they are aligned with natural systems. One thing is sure—it will not be Nature's laws, the delicate interdependence that characterizes Earth's ecosystems, that changes. The change we seek must begin within each of us, and from there permeate throughout our social, political, and economic structures.

A few numbers quoted at the end of *Home,* a documentary by Yann Arthus-Bertrand (2009) that shows stunning and colorful aerial views of the Earth, can serve as a wakeup call:

- 20 percent of the world's population consumes 80 percent of its resources
- 5000 people a day die because of unsafe drinking water
- One billion people have no access to safe drinking water
- Over 50 percent of grain traded around the world is used for animal feed or biofuels
- Three quarters of fishing grounds are exhausted, depleted or in dangerous decline
- Nearly one billion people are going hungry
- The ice cap is 40 percent thinner than 40 years ago

While a few still doubt that our survival is at stake, most people know that something needs to be done. The question is, can we change fast enough? If global warming continues unabated, the permafrost will melt, releasing methane into the atmosphere, and the sea levels will rise ever faster, with catastrophic consequences. Many scientists now say that we have a mere ten years before climate change hits a tipping point and the Earth's climate systems begin to oscillate wildly. Such climate instability would undermine the agricultural foundations of modern society and force millions of climate refugees to migrate in search of higher ground, food, and clean water—sparking countless regional conflicts over dwindling resources.

Even when confronted with such frightening prospects, we know that transformation is possible and that many people are willing, even yearning, to change. President Obama was elected based on his campaign platform of *change*. Since taking office, he has made progressive declarations concerning nuclear weapons, dialogue of civilizations, human rights, sustainability and other topics that no one would have believed possible, until recently, for someone in his position.

But what is it, precisely, that we need to change? How do we change our behavior, and in which direction? Where do we start? In her chapter, Marjorie Kelly points to the human spirit as the most crucial point of departure towards all transformations. She also captures the spirit of urgency pervading this entire volume:

> What we are experiencing today is a systemic crisis, heralding the beginning of a large-scale transformation of the deepest structures of cultural organization. In systems thinking, transformation is distinct from adaptation, which is the normal process of incremental adjustments in policies and institutions. Transformations are something else. They are those rare moments in history when dominant institutions become unfit for an unfolding new set of circumstances and change in profound ways.
>
> We're experiencing today a system event operating at the scale of the planet. We are on the threshold of a passage likely to prove as significant as the advent of agriculture and the dawn of the Industrial Revolution. [...] We are poised on the narrow edge of the present and will ultimately tip into one of two likely futures.
>
> One is a kind of Fortress World, a landscape of devastation where the affluent come to live in protected enclaves amid oceans of misery. The other is a world of Great Transition. [...] The critical difference between a Fortress World and a Great Transition is one of cultural spirit. A spirit of fear. Or a spirit of hope. It is our spirit that will allow us to transform the inevitable crises ahead in positive ways. Which scenario unfolds will depend upon the cultural ethos we both create and inhabit: an ethos that says each of us is out for ourselves, or we're all in this together.

The contributors to this volume all believe that change is urgently needed and possible, and they offer specific examples. They suggest solutions that demand new perspectives in our ethics, beliefs, traditions, economy, business, gender relations, education, and technology. They show that we need to, and have the power to, implement changes in all of these areas.

Our authors each have something special to offer. David Abram is a gifted writer, and director of the Alliance for Wild Ethics, whose book, *The Spell of the Sensuous* (1997) has received the international Lannan Literary Award for nonfiction. Marjorie Kelly was heralded as the new Thomas Paine for her seminal work in the area of sustainable business systems, *The Divine Right of Capital* (2001). Dharm Bhawuk's co-authored paper, published in this volume, won the Rupe Chilsom Practical Theory Best Paper Award from the Organization Development and Change Division of the Academy of Management in 2009. Deva Temple, a promising young author, scholar, and activist (who has also served as co-editor of this book) offers us a fresh look at the role of gender relations in shaping how we interact with one another and with the natural world. Ellen Gaddy is co-director of NewDay Africa and graduate student of psychology and social transformation at Saybrook University. Her areas of focus encompass the psychology of sustainability as well as monetary transformation and reform. Imre Lévai is a notable scholar who serves as scientific advisor at the Institute for Political Science of the Hungarian Academy of Sciences. Robin Attfield has taught philosophy since 1968 and currently serves as professorial research fellow at Cardiff University. Also, in trying to break out from the tyranny of monolingualism, the Toda Institute is proud to introduce the pioneering research of Izaskun Petralanda and Laura Delgado Petrocelli, two dedicated and accomplished scholars from Venezuela. The effort required to prepare academic papers in a language not one's own is deeply appreciated.

The original idea for this volume was born during the Toda Conference held in Honolulu in November 2008, entitled "Facing Climate Change with a Renewed Environmental Ethics." This book represents the contributions that address not only climate change, but also many other areas where human behavior needs to change urgently if we are to make our lives on this planet sustainable. At the end of the Climate Change Conference, all participants agreed on a common text, entitled "The 2008 Toda Declaration on Climate Change."

The nine chapters in this book have been organized around three keywords: "Reconnecting, Reframing, and Rethinking." We need to *reconnect* with the Earth and nature, as well as with each other and parts of ourselves that we have ignored for too long. We need to *reframe* the way we prioritize choices in our economy, the way we do business and pursue development. We need to *rethink* the mission of education and the roles of technology.

Reconnecting with the Earth and Humanity

With "Storytelling and Wonder," David Abram opens this volume with a strong recommendation that we reconnect with oral culture. Indeed, *Literate Culture*, with its books, newspapers, and journals may be cosmopolitan and highly portable, but words on paper are only abstractions. *Computer Literacy* is global and globalizing, and allows us to explore any place on Earth virtually. However, we only experience the world through the keyboard, and connect with disembodied electronic fields of information. *Oral Culture* is, by contrast, local. The tales shared by storytellers are about the places we inhabit, and these stories connect us with the concrete, real-life landscapes where the narrative takes place. With oral communication, we can also use our breathing bodies and feel the vibrant voice of others. Abram does not recommend abandoning books and computers, but rather that we become aware once again of the potential of oral traditions that are at the source of our living cultures. A deeper appreciation for oral culture around the world would create a profound change in the human spirit.

With "The Ecology of Equality," Deva Temple recommends a double movement of our spirit towards more sustainability. First, she suggests a rebalancing of ethics, away from the traditionally masculine ethos of dominance and toward a more traditionally feminine ethos of shared sacredness and interdependence. The desire to rule over and compete for control is seen here as the basic foundation of our economic system and as the ideological and sociological soil from which great technological advances have been made, advances which have brought both countless benefits and untold hardships to humankind, and which are now threatening to unravel the web of life on Earth. We need to shift the emphasis from dominance toward shared sacredness, with an enhanced awareness of interdependence, and with a spirit of nurturing and compassion. However this is only one half of the equation. Temple then challenges the very conceptual framework that separates male and female. The notion that there are two separate ways of being, one for men, characterized by dominance, and the other one for women, characterized by nurturing, is challenged. Actually, Temple argues, both worlds coexist inside each person, and what is necessary is to bring out the best in ourselves at the right time, be it the courageous capacity to lead, or the compassionate ability to nurture.

Finally, Robin Attfield concludes this section with "Exploring Ethical and Religious Attitudes to Sustainability" a chapter exploring several philosophical approaches to sustainability, including that of philosophers John Rawls, Immanuel Kant, Aristotle, and John Stuart Mill. He concludes that there are loopholes in each philosophical approach to sustainability and then refocuses on religious approaches to sustainability, advocating for the biblical concept of stewardship as a key principle that might inspire a new religious-secular consensus towards sustainability.

Reframing Economy and Development

The second section begins with a chapter by Ellen Gaddy, graduate student of psychology and social transformation at Saybrook University and co-director of NewDay Africa, who provides us with insights concerning the need to change our attitude towards money, and for the sustainable economists to collaborate with psychologists, the peace movement, and other actors in order to actualize sustainable abundance. In "Against So Much Money and Power Can the Sustainability Movement Succeed?" Gaddy explores the structure and function of the centralized monetary system and points out the ways in which this system creates a growth imperative that exacerbates resource depletion and waste creation. She argues that changes to the monetary system are integral to any attempts to create a more sustainable future and explores the benefits and limitations of several possible ways to reform monetary policy that could provide a basis for a sustainable economic system.

Marjorie Kelly in "Economic Genesis" encourages us to reframe the social architecture of our economy. Our core social institutions are corporations and financial markets, which have lead us to unsustainable ways of life. What kinds of values should be at the core of our philosophies, our economies, and our societies? Kelly presents us with a choice between *self-interest, economic growth, and free markets* on the one hand, and *interdependence, sustainability, and well-being* on the other. She argues that we must choose the latter if we are to avoid the devastating consequences of what she refers to as the *Fortress World*, a world in which the wealthy live in increasingly isolated and vulnerable fortresses, while the masses struggle to survive and society crumbles. This hope-filled work helps us to see neither optimistically, nor pessimistically, but rather realistically, and with a dogged determination to make the right decision.

Our traditional economic models imply standardized ways of doing business and inducing development. In "From Social Engineering to Community Transformation," Dharm Bhawuk, Susan Mrazek, and Vijayan Munusamy offer three examples of successful ways to implement new paradigms in social science research and development programs. Three corporations—Amul, the Grameen Bank, and Mondragon—are used as examples to show the validity of the theories of community psychology presented here. The authors recommend essential approaches to effective community transformation that are based on a philosophy of nurturing and empowering people, rather than solving problems for them through legislations or rules.

Rethinking Education and Technology

The final section begins with a chapter, "Sustainable Globalization and Limits to Sustainability," by Imre Lévai, who argues that the current academic paradigms and models which underlie sustainability theory and practice are overly simplistic and in need of reconsideration to take into account the non-linear and complex nature of the world system. This piece represents one strain of academic thinking and reviews sustainability through the lens of Immanuel Wallerstein's world systems theory, applying a critical analysis to the work of the Club of Rome in developing theoretical foundations for sustainability. He concludes by arguing for radical shifts in academic discourse and theoretical constructs which underpin the theory and practice of sustainability.

In "Renewal of Ecological Ethics," Izaskun Petralanda draws lessons from the Amazonian Aboriginal cultures. Whereas our Western materialistic and globalizing society encourages the individual to be successful as "being-person-*in*-the-world," the perspective of Amazonian Aboriginal cultures is that we are all "being-person-*with*-the-world," leading to the concept of "environmental solidarity." Indeed, it is quite possible that our "human ability to relate" has been impaired by ethical systems that emphasize individuality over interdependence. What kind of relations are we engaging in with ourselves, other people, other living beings, and nature? We might entertain the idea that we are simply a part of nature, but for the Yanomami People, the forest and its animate and inanimate inhabitants are all a part of us as well. Petralanda suggests the importance of placing ecoethics at the center of education, allowing people to develop a new awareness of an ancient perspective that she translates in a poetic and compelling way as "being-person-careful-caring-and-loving-with-the-world."

Concluding this volume, Laura Delgado Petrocelli, in "The Case of Malaria in Sucre, Venezuela," shows how technology, rather than being the main cause of our destructive power, can be seen as a conduit for the employment of renewed ethical values. Using the most advanced research tools, Delgado was able to show patterns in the spread of malaria in the Sucre region of Venezuela. Based on a capacity to think in systemic and holistic ways, one can recognize that malaria relentlessly progresses wherever people destroy the environment through degrading land use patterns. This discovery, which was accelerated through the use of technology, shows that we need to be keenly aware of interdependence. We might think that land use is "good," providing work and revenues, with no connection to malaria, but actually both are deeply linked—as are all things. Here we can see how cutting-edge technology can encourage the type of vision of "being-person-with-the-world" recommended by the Amazonian Aboriginal cultures in the previous chapter.

This short summary shows that transformation is possible, and can be seen as a fascinating adventure rather than an overwhelming and anxiety-producing task. Whatever field of activity we are involved in, we have in our power the capacity to bring about fundamental transformations here and now. These examples of reconnecting, reframing and rethinking are only illustrations of the potential for positive transformation that lies ahead.

The Toda Institute is delighted to be able to present this volume at such a crucial time in the history of our planet. We will leave you with this striking message, presented at the end of the documentary *Home* mentioned above: "It is too late to be pessimists."

Olivier Urbain
Tokyo, Japan
June 2009

Storytelling and Wonder: On the Rejuvenation of Oral Culture as an Ecological Imperative

David Abram

Introduction

In the prosperous land where I live, a mysterious task is underway to invigorate the minds of the populace, and to vitalize the spirits of our children. For a decade now, parents, politicians, and educators of all forms have been raising funds to bring computers into every household in the realm, and into every classroom from kindergarten on up through college. With the new technology, it is hoped, children will learn to read much more efficiently, and will exercise their intelligence in rich new ways. Interacting with the wealth of information available online, children's minds will be able to develop and explore much more vigorously than was possible in earlier eras—and so, it is hoped, they will be well prepared for the technological future.

How can any child resist such a glad initiative? Indeed, few *adults* can resist the dazzle of the digital screen, with its instantaneous access to everywhere, its treasure-trove of virtual amusements, and its swift capacity to locate any piece of knowledge we desire. And why *should* we resist? Digital technology is transforming every field of human endeavor, and it promises to broaden the capabilities of the human intellect far beyond its current reach. Small wonder that we wish to open and extend this powerful dream to all our children!

It is possible, however, that we are making a grave mistake in our rush to wire every classroom, and to bring our children online as soon as possible. Our excitement about the Internet should not blind us to the fact that the astonishing linguistic and intellectual capacity of the human brain did not evolve in relation to the computer! Nor, of course, did it evolve in relation to the written word. Rather it evolved in relation to orally told stories. Indeed, we humans were telling each other stories for many, many millennia before we ever began writing our words down—whether on the page or on the screen.

The Power of Spoken Story

Spoken stories were the living encyclopedias of our oral ancestors, dynamic and lyrical compendiums of practical knowledge. Oral tales told on special occasions carried the secrets of how to orient oneself in the local cosmos. Hidden in the layered adventures of their characters were precise instructions for the hunting of various animals, and for enacting the appropriate rituals of respect and gratitude if the hunt was successful, as well as specific insights regarding which plants were good to eat and which were poisonous, and how to prepare certain herbs to heal cramps, or sleeplessness, or a fever. The stories carried instructions about how to construct a winter shelter, and what to do during a drought, and—more generally—how to live well in this land without destroying the land's wild vitality.

Such practical intelligence, intimately related to a particular place, is the hallmark of any deeply oral culture. Continually tested in interaction with the living land, altering in tandem with subtle changes in the local earth, even today such living knowledge resists the fixity and permanence of the printed page. Because it is specific to the way things happen *here,* in this high desert—or coastal estuary, or mountain valley—this kind of intimate intelligence loses its meaning when abstracted from its terrain, and from the particular persons and practices that are a part of its terrain. Such place-specific savvy, which deepens its value when honed and tempered over the course of several generations, forfeits much of its power when uprooted from the soil of its home and carried—via the printed page or the glowing screen—to other places. Such intelligence, properly speaking, is an attribute of the living land itself; it thrives only in the direct, face-to-face exchange between those who dwell and work in this place.

So much earthly savvy was carried in the old tales! And since, for our indigenous ancestors, there was no written medium in which to record and preserve the stories—since there were no written books—the surrounding landscape, itself, functioned as the primary *mnemonic,* or memory trigger, for preserving the oral tales. To this end, diverse animals common to the local earth figured as prominent characters within the oral stories—whether as teachers or tricksters, as buffoons or as bearers of wisdom. Hence, a chance encounter with a particular creature as a tribesman went about his daily business (an encounter with a coyote, perhaps, or a magpie) would likely stir the memory of one or another story in which that animal played a decisive role. Moreover, crucial events in the stories were commonly associated with particular *sites* in the local terrain where those events were assumed to have happened; whenever one noticed that place in the course of one's daily wanderings—when one came upon that particular cluster of boulders, or that sharp bend in the river—the encounter would spark the memory of the storied events that had unfolded there.

Thus, while the accumulated knowledge of our oral ancestors was carried in stories, *the stories themselves were carried by the surrounding earth.* The local landscape was alive with stories! Traveling through the terrain, one felt teachings and secrets sprouting from every nook and knoll, lurking under the rocks and waiting to swoop down from the trees. The wooden planks of one's old house would laugh and whine, now and then, when the wind leaned hard against them, and whispered wishes would pour from the windswept grasses. To the members of a traditionally oral culture, all things had the power of speech.

Indeed, when we consult indigenous, oral peoples from around the world, we commonly discover that for them there is no phenomenon—no stone, no mountain, no human artifact—that is definitively inert or inanimate. Each thing has its own spontaneity, its own interior animation, its own life! Rivers *feel* the presence of the fish that swim within them. A large boulder, its surface spreading with crinkly red and gray lichens, is able to influence the events around it, and even to influence the thoughts of those persons who lean against it—lending their reflections a certain gravity, and a kind of stony wisdom. Particular fish, as well, are bearers of wisdom, gifting their insights to those who catch them. Everything is alive—even the stories themselves are animate beings! Among the Cree of Manitoba, for instance, it is said that the stories, when they are not being told, live off in their own villages, where they go about their own lives. Every now and then, however, a story will leave its village and go hunting for a person to inhabit. That person will abruptly be possessed by the story, and will soon find herself telling the tale out into the world, singing it back into active circulation....

There is something about this storied way of speaking—this acknowledgement of a world all alive, awake, and aware—that brings us close to our senses and to the palpable, sensuous world that materially surrounds us. Our animal senses know nothing of the mechanical, quantifiable world to which so much of our civilized discourse refers. Wild and gregarious organs, our senses spontaneously experience the world not as a conglomeration of inert objects but as a field of animate presences that actively *call* our attention, that *grab* our focus or *capture* our gaze. Whenever we slip beneath the abstract assumptions of the modern world, we find ourselves drawn into relationship with a diversity of beings as inscrutable and unfathomable as ourselves. Direct, sensory perception is inherently animistic, disclosing a world wherein every phenomenon has its own active agency and power.

When we speak of the earthly things around us as determinate objects or passive "natural resources," we contradict our spontaneous sensory experience of the world, and hence our senses begin to wither and grow dim. We find ourselves living more and more in our heads, adrift in a sea of abstractions, unable to feel at home in an objectified landscape that seems alien to our dreams and emotions. But when we begin to tell stories, our imagination begins to flow out through our eyes, through our ears, to inhabit the breathing earth once again. Suddenly, the trees along the street are looking at us, and the clouds crouch low over the city as though they are trying to hatch something wondrous. We find ourselves back inside the same world that the squirrels and the spiders inhabit, along with the deer stealthily munching the last plants in our garden, and the wild geese honking overhead as they flap south for the winter. Linear time falls away, and we find ourselves held, once again, in the vast cycles of the cosmos—the round dance of the seasons, the sun climbing out of the ground each morning and slipping down into the earth every evening, the opening and closing of the lunar eye whose full gaze attracts the tidal waters within and all around us.

For we are born of this animate earth, and our sensitive flesh is simply our part of the dreaming body of the world. However much we may obscure this ancestral affinity, we cannot erase it, and the persistence of the old stories is the continuance of a way of speaking that blesses the sentience of things, binding our thoughts back into the depths of an imagination much vaster than our own. To live in a storied world is to know that intelligence is not an exclusively human faculty located somewhere inside our skulls, but is rather a power of the animate earth itself, in which we humans, along with the hawks and the thrumming frogs, all participate. It is to know, further, that each land, each watershed, each community of plants and animals and soils, has its particular *style* of intelligence, its unique mind or imagination evident in the particular patterns that play out there, in the living stories that unfold in that valley, and that are told and retold by the people of that place. Each ecology has its own psyche, and the local people bind their imaginations to the psyche of the place by letting the land dream its tales through them.

Vernacular Culture and the Primacy of Place

Today, economic globalization is rapidly undermining rural economies and tearing apart rural communities. The spreading monoculture degrades both cultural and biotic diversity, forcing the depletion of soils and the wreckage of innumerable ecosystems. As the civilization of total commerce muscles its way into every corner of the planet, countless species tumble helter-skelter over the brink of extinction, while the biosphere itself shivers into a bone-rattling fever.

For like any living being, earth's metabolism depends upon the integrated functioning of many different organs, or ecosystems. The human body could not possibly maintain its health if the lungs were forced to behave like the stomach, or if the kidneys were forced to act like the ears or the soles of the feet. Likewise, the planet's metabolism is thrown into disarray when each region is compelled to behave like every other region—when diverse places and cultures are forced to operate according to a single, mechanical logic, as interchangeable parts of an undifferentiated, homogenous sphere.

In the face of the ever-expanding monoculture and its technological imperatives, more and more people are coming each day to recognize the critical importance of revitalizing local, face-to-face community. They recognize their common embedment within the life of this breathing planet, yet they know that such unity arises only from a vital and thriving multiplicity. A reciprocal respect and interdependence between richly *different* cultures—each a dynamic expression of the unique earthly place, or bioregion, that supports it—is far more sustainable than a homogenous, planetary civilization.

Many of us have already worked for several decades on ecological and bioregional initiatives aimed at renewing local economies and the conviviality of place-based communities. Yet far too little progress was made by the movements for local self-sufficiency and sustainability. To be sure, our efforts were hindered by the steady growth of an industrial economy powered by the profligate burning of fossil fuel. Yet our efficacy was also weakened by our inability to recognize the immense influence of everyday language. Our work was hobbled, that is, by our inability to discern that the spreading technologization of everyday life in the modern world (including the growing ubiquity of automobiles, telephones, and televisions, and more recently of personal computers, cell phones, and palmtop digital assistants) had been accompanied by a steady transformation in language—by an increasing abstractness and generality in daily discourse. Local vernaculars had fallen into disuse; local stories had been forgotten; the oral forms and traditions by which place-specific knowledge had long been preserved and disseminated were no longer operative.

We at the Alliance for Wild Ethics (AWE) now recognize that a rejuvenation of real, face-to-face community—and the sensorial attunement to the local earth that ensures the vitality and sustenance of such community—simply cannot happen without a rejuvenation of the layer of language that goes hand in hand with such attunement. It cannot happen without renewing that primary layer of language, and culture, that underlies all our more abstract and technological forms of discourse. A renewal of place-based community cannot happen without a renewal of oral culture.

Mediated Communication and the Air (the Original Medium)

Every medium of communication organizes our awareness in a particular way, each engaging us in a particular form of community. Without here analyzing all the diverse media that exert their claims upon our attention, we can acknowledge some very general traits:

- Literate culture (the form of society implicitly informed by books, newspapers, and magazines) enables an exchange of tales and insights drawn from different places and different eras. Reading is a wondrous mode of experience, yet it is necessarily abstract relative to our direct, sensory encounters in the immediacy of our locale. Indeed, by mingling notions originally rooted in divergent cultures, literate discourse instills a useful distance, and detachment, from our immediate environment. In this sense, literate culture is inherently *cosmopolitan.*
- Computer literacy, and our engagement with the Internet, brings us almost instantaneous information from around the world, empowering virtual interactions with people from vastly different communities. Yet such digital engagements are even more disembodied and placeless than our involvement with printed books and magazines. Cyberspace seems to have no location at all, unless the "place" that we encounter through the Internet is, well, the planet itself, transmuted into a weightless field of information. In truth, our increasing participation with email, e-commerce, and electronic information involves us in a discourse that is inherently *global* and *globalizing.* (It is this computerized form of communication, of course, that has enabled the rapid globalization of the free-market economy.)
- Oral culture (the culture of face-to-face storytelling) is inherently *local.* Far more concrete than those other modes of discourse, genuinely oral culture binds

> us not only to our immediate human community, but to the more-than-human community—the particular ecology of animals, plants and earthly elements in which we materially participate.

In contrast to more abstract forms of media, the primary *medium* of oral communication is the atmosphere itself. Spoken words, after all, are shaped breath. We inhale this invisible element, circulating it within lungs, and as we exhale we allow our breath to vibrate two membranes within our throat, shaping the breath with our tongue and lips and palate as we sound our stories or songs into the world. Air, which is subtly different in each terrain, and which binds our own breathing bodies to the metabolism of oak trees and hawks and the storm clouds gathering above the city, is the implicit intermediary in all oral communication. As the most ancient and longstanding form of human discourse, oral culture provides the necessary soil and sustenance for those more abstract styles of communication and reflection.

Replenishing Oral Culture

A revitalization of oral culture does not entail that we renounce reading and writing, nor that we cease interacting with our computers. It entails only that we leave space in our days for an interchange with one another and with the earth that is *not* mediated by technology—neither by the television, nor the computer, nor even the printed page.

Among writers, for instance, it entails that we allow that there are certain stories we might come upon that should *not* be written down—stories that we instead begin to tell, with our own tongue, in the particular terrain where those stories live.

For parents and grandparents, it entails that we set aside, now and then, the storybooks that we read to our children in order to actually *enact* a story *with the whole of our gesturing body*—or better yet, that we draw our kids out of doors in order to improvise a tale about the wild wind that's now blustering its way through these city streets, plucking the hats off people's heads....

Among educators, it entails that we begin to rejuvenate the arts of telling, and of listening, in the context of the living landscape where our lessons happen. For too long we have incarcerated the potent magic of linguistic meaning within an exclusively human space of signs. Hence the land itself has fallen mute; it now seems little more than a passive backdrop for human events, or a storehouse of resources waiting to be mined for purely human purposes. Can we return to the local land an implicit sense of its own meaning and eloquence? Not without renewing the sensory craft of listening, and the sensuous art of storytelling. Can we help our students to translate the quantified abstractions of science into the language of direct experience, so that those necessary insights begin to come alive in our felt encounters with the animate earth around us? Can we begin to affirm our own co-evolved, carnal embedment within this blooming, buzzing proliferation of life, stirring a new humility in the face of a world that we did not create—in the face of a world *that created us?* Most importantly, can we begin with our students to *restore* the health and integrity of the local earth? Not without *restorying* the local earth. For our senses have become exceedingly estranged from the earthly sensuous. The age-old reciprocity between the human animal and the animate earth has long been short-circuited by our increasing involvement with our own creations, our own human-made technologies. And yet a simple tale, well told, can shatter the spell—whether for an hour, or a day, or even a lifetime.

There is no need to give up reading, nor to discard our computers, as long as we recall that such mediated and technological forms of interchange inevitably remain rooted in the more primary world of direct experience. As long as we remember, that is, that our involvements with the printed page and the digital screen draw their basic sustenance from our more immediate, face-to-face encounter with the flesh of the real.

Conclusion

The Alliance for Wild Ethics holds that the globalizing culture of the Internet, and the cosmopolitan culture of books, are both dependent, for their integrity, upon the place-based, oral culture of face-to-face storytelling. Oral culture – and the vernacular intimacy with the local land that goes hand-in-hand with such culture – is the forgotten ground that still supports those more abstract layers of communication. It is the neglected but necessary soil from whence civilization still draws its sustenance, the nourishing humus in which our humanity remains rooted.

Left to itself the literate mind, adrift in the play of signs, comes to view nature as a sign, or a complex of signs. It forgets that the land is not first and foremost an arcane text to be read, but a community of living, speaking beings to whom we are beholden. Adept at *representing* the world verbally, the literate intellect forgets how to orient in the midst of the world's *presence*, how to hear those many voices that do not speak in words. Similarly the computerized mind, when left to its own devices, all too easily overlooks the solid things of the earth. Skilled in the rapid manipulation of symbols, it neglects the stones and the grasses that symbolize nothing other than *themselves.* Dazzled by its own virtual creations, the digital self forgets its dependence upon a world that it did not create, overlooking its carnal emplacement in the very world that *created it.*

When oral culture degrades, then the literate mind loses its bearings, forgetting its ongoing debt to the body and the breathing earth. When stories are no longer being told in the woods or along the banks of meandering rivers—when the land is no longer being honored, *ALOUD,* as an animate, expressive power—then the human senses lose their attunement to the surrounding terrain. We no longer feel the particular pulse of our place—we no longer hear, or respond to, the many-voiced eloquence of the land. Increasingly blind and deaf, increasingly impervious to the sensuous world, the technological mind begins to lay waste to the earth.

We can be ardent readers (and even writers) of books, and enthusiastic participants in the World Wide Web and the Internet, while recognizing that these abstract and almost exclusively human layers of culture will never be sufficient unto themselves. Without rejecting these rich forms of communication, we can nonetheless discern, today, that *the rejuvenation of oral culture is necessary to a sustainable and sustaining world.*

The Ecology of Equality: From Dominance to Shared Sacredness

Deva Temple

Crossroads

Humanity, as a species, is at a crossroads. In the last two centuries, technological advances have allowed the human population to increase exponentially, from 1 billion in 1776 to 6.5 billion in 2006.[1] The same technologies have also brought unprecedented, consumption-based, material wealth to an elite handful of people,[2] while simultaneously unraveling our planet's ability to sustain life. Humans are utilizing resources at an ever-increasing rate, perpetually obsessed with next quarter's financial statements, single-mindedly determined to develop underdeveloped countries, even if that process requires more resources than the earth can produce. As a result, humanity will soon face not only the effects of climate change, but also massive global food shortages, intense drought, rising sea levels, loss of farmland to soil salinization and desertification, mass migration of ecological refugees, loss of biodiversity due to increased species extinction, the collapse of ocean fisheries, increased occurrence of preventable and communicable diseases, global economic depression, and increased levels of conflict.[3]

These challenges to life on planet Earth supersede the constraining boxes of ideology and politics. As Al Gore (2006) reminds us, climate change is not only a political issue but also a *moral* issue. The notion that humans have a moral or ethical responsibility to care for the planet may be disjointing for those of us who have spent our lives in political systems that are based on dominance, economic systems based on greed (otherwise known as "rational self-interest"), and religious traditions that devalue the material world in search of the transcendent. How can we care for the Earth, on which our very existence depends, while our hearts and minds are burdened with ideas of superiority, mastership, or exploitation? If we are to survive as a species, we will have to alter not merely our technology, or our political and economic systems, but the very foundations of our cultural and ethical paradigms. We must balance the traditionally masculine ethics of dominance, which has guided our technological and economic advances, with a more feminine ethos of shared sacredness that promotes respect for our fellow human beings and nature alike.

Some caveats are necessary before delving deeply into any discussion of gender. In discussions of the "feminine" and the "masculine" it is important to recognize that the prevalent archetypes of feminine and masculine are dynamic and vary throughout time, across cultures, and among individuals. Also, it should be mentioned that the roots of such archetypes continue to be a source of debate as demonstrated by the ongoing and hoary nature-nurture argument.[4] This chapter is not focused on the traits of individuals nor is it assumed that these archetypes are an accurate and authentic manifestation of either the true feminine or the true masculine. The ways in which our biology impacts our psychology, and the ways in which our cultures shape our personalities, are hard to empirically establish; no attempt is made to do so here.

It must also be stated that when speaking of *feminine/masculine* ethics and *feminine/masculine* ways of being, one is actually talking about *human* ethics and *human* ways of being, not only because women and men are human, but because these ethics and ways of being are available to, and exist within, all of humanity regardless of gender. It is accurate to say that women are capable of rage and that men are capable of compassion. Both extremes exist within each and every human. According to Bakan (1966), all living organisms can be characterized as exhibiting two basic ways of being, *agency* and *communion.* Bakan defined agency as a sense of self that is manifested through self-assertion, self-enhancement, and self-protection and was thought to be masculine. Communion entailed selflessness, concern for others, and group participation, and was considered a feminine quality. We must be brave enough to reach beyond these proscribed boxes and to seek an authentic self. Indeed, this possibility is one of feminism's greatest gifts to the world; we are no longer bound to live our lives in narrowly limited, externally defined ways simply because of our physical appearance.

The focus of Second Wave feminism was to free women to adopt traditionally masculine ways of being, as well as the accompanying ethics of dominance, hierarchy, and competition. Because feminine ways of being had been historically denigrated, women felt that the only way to gain recognition as human beings was to move out into the world and claim masculine space, occupations, and activities. This may have been a requisite step, as women seek to attain a place of power and respect from which to articulate feminine ethics, but there has not been a parallel men's movement that has encouraged men to take up feminine activities and ethics (Farrell, 1994). This imbalance may stem from the differential ascribed value of masculinity and femininity,—to be masculine is more prestigious than to be feminine,—but the result is that women have achieved greater individual freedom, while men are trapped in an increasingly confusing double-bind. For men, to be masculine in the traditionally dominance-dependent manner has become politically incorrect, while to be traditionally feminine in any manner is shameful and taboo. What is necessary is a reconciliation of these seemingly opposing value systems, and ways of being, and a freeing of individuals to experiment, to embrace ethics and ways of being which are fluid, dynamic, and authentic expressions of the self. The realization of individual freedom may be but a means to a more important end, the evolution of human morality, consciousness, and the continued existence of our species (Aptheker, 2003).

The Disproportionate Prevalence of Dominance Ethics

Dominance is the supreme value of Western culture (French, 1985; Greene, 1998; Machiavelli, 1513). According to the *Merriam Webster Dictionary*, the definition of dominant is "commanding, controlling, or prevailing over all others … very important, powerful, or successful … overlooking and commanding from a superior position." This definition of dominance has been a defining principle of masculinity, as men are expected to seek dominant positions in society and in relationships with women, other men, and nature (Farrell, 1993; Goldberg, 1993). Explanations of male dominance range from the biochemical (Goldberg, 1993) to the socially constructed (French, 1985), but there is consensus around the notion that to be masculine requires dominance seeking behavior and to be dominant is to adopt a masculine way of being, whether or not one is physically male.

Not only is the pursuit of dominance an indicator of masculinity in individuals, but the competitive nature of dominance seeking behavior is also characteristic of masculine cultures (Hofstede, 2001). In masculine cultures the accumulation of money and material possessions is of primary importance and the health of relationships is of lesser importance (Hofstede, 2001, 1998). Likewise, economic growth is more important than environmental protection and conflicts are more likely to be solved by force than by negotiation in masculine cultures (Hofstede, 2001).

In Western culture, the preference for masculine values has a long history. According to Aristotle, "The male and the female differ from each other in the possession of an ability and in the lack of an ability. The male is able to concoct, formulate and to ejaculate the sperm which contains the *origin of the form* [or the transcendent, defining quality of the being]—I do not mean here the material element out of which it is born resembling its parent but the initiating formative principle whether it acts within itself or within another. The female, on the other hand, is that which receives the seed [soil] but is unable to formulate or to ejaculate it" (765 b9-16). While knowledge of the genetic contributions of females was, at the time of Aristotle's writing, a long way off, it was believed that the male form alone possessed the ability to plant the seed of new life, while the female merely acted as a passive vessel, providing raw materials for the new being, but not determining its individual form or qualities. Such a conceptualization was used to justify various manifestations of patriarchy (Lerner, 1986).

The disproportionate importance placed on the transcendent and determining power of masculinity is repeated in the way in which our economic system values the material contributions of nature.

> For all the discussions on value added, one would think that there would be some discussion of *that to which value is being added.* But modern economists say no more about it.... It is just "matter," and its properties are not very interesting.... That to which value is added is therefore inert, undifferentiated, interchangeable, and superabundant—very dull stuff indeed, compared to the value-adding agents of labor, with all its human capacities, and capital, which embodies the marvels of human knowledge (Daly, 1996: 62-63).

Kelly (2003) takes this a step further, pointing out that our basic accounting equations place payments made to labor on the cost side and payments made to shareholders on the profit side, which disconnects the productive relationship of each party to the profits being earned. While laborers contribute the value added to the "matter" of nature, investors may never even visit the factory and workplaces where such work is being done—yet the profits are deposited into their accounts. This odd reversal of valuation creates a hierarchy of dominance in which humans dominate, and are more valuable than non-human nature, and *certain* humans dominate and are more valuable than others (as Marjorie Kelly states in her chapter in this present volume). Those humans who work primarily with their bodies are valued less than those who work with their minds. Above all, commanding from on high, are those humans who *do not work* (Domhoff, 2006). We call them "investors" and credit them with providing "capital ... the marvels of human knowledge" (Daly, 1996: 63).

While human knowledge is elevated in importance, the body is a source of shame, to be denied and controlled (Hite, 1994). Descartes famously proved his very existence with reference to his mind, which he believed was the eternal and true human identity, "Cogito, ergo sum," "I think therefore I am." Descartes established a philosophy that completely (and falsely) separated mind from matter, *res cogitans,* "thinking thing" from *res extensa,* "extended thing" (Capra, 1982). The mind had value, the matter, not so much. This Cartesian view of the relative value of the mind and body, mere matter, has been of seminal importance to Western culture (Capra, 1982).

It is interesting to note that the word matter is derived from the word *mater*, meaning "origin, source, mother" (Harper, 2001). The devaluation of human mothers and of "Mother Earth" is related. In fact, it was Sir Francis Bacon who best articulated the common antipathy of man toward woman and toward nature, "Nature, in his view, had to be 'hounded in her wanderings,' 'bound into service,' and made a 'slave.' She was to be 'put in constraint,' and the aim of the scientist was to 'torture nature's secrets from her'" (Capra, 1982). Is it any wonder that a technologically advanced civilization with such a paradigm would destroy its planet without concern? Our continued acceptance of the ethics of dominance is reflected in our military, economic, and political institutions (Mills, 1956), and is at the core of societal patterns that threaten the environment.

The Role of Power in the Construction of Dominance Hierarchies

Dominance is achieved, and hierarchies are created, through power plays that have deep roots in Western culture. The creation story in Genesis reflects this: Adam and Eve are punished for their disobedience, for their desire to "be like God, knowing good and evil" (Gen. 3:5, New International Version). Their punishment for seeking forbidden knowledge, or power, was death, pain, enmity, *male dominance,* and *female submission*. God continues on to express anxiety that, "The man has now become like one of us, knowing good and evil. He must not be allowed to reach out his hand and take also from the tree of life and eat, and live forever" (Gen. 3:22). God feels threatened that humans might become god-like through the acquisition of knowledge and eternal life. This interaction between the powerful and the powerless, but *potentially* powerful, is emblematic of subsequent power relations, not only between members of the ruling class (who, during the Dark Ages, ruled by divine right), but between the serpent and humans. The serpent is cursed to "crawl on [its] belly and ... eat dust" (Gen. 3:14), in other words, the serpent was punished with proximity to the dirt, the ground, to nature, to the Earth.

In his national bestseller, *The 48 Laws of Power*, Green (1998) explicitly shares several approaches to securing hierarchical power. Among them: "Conceal your intentions ... get others to do the work for you, but take all the credit ... keep people dependent on you ... crush your enemy totally ... keep others in suspended terror ... do not commit to anyone ... control the options ... get others to play with the cards you deal ... Pose as a friend, work as a spy ... If there is something you want but cannot have, show contempt for it," etc. ... (Green 1998, *ix-xvi*). What Green describes is the mindset of Medieval European courtiers as they vied for power in the royal courts. This form of power is achieved through deceit and its goal is a certain kind of power, the power-over. It is the kind of power that allows you to get someone else to do what you want them to do. Indeed, Machiavelli (1513) wrote about this kind of power in *The Prince* several centuries ago. Western culture is obsessed with power-over and it is by no means limited to our political institutions.

This brings us to a final point about the nature of power. There are two kinds of power, power-over and the power-to (Allen, 2008). The man digging the ditch has the power-to, the man supervising him has the power-over. The woman giving birth has the power-to, the father bestowing his last name upon the child has the power-over. We value only one face of power: power-over. In contrast, power-to is almost scorned. It requires that one "lower" oneself to interact with nature, to become like nature, to become, oneself, productive. Only those who are able to cast themselves as removed from nature, and from the natural toil of the creative process, can appear to wield power-over. Those who live in direct and authentic relationship with nature, who survive and create through their interdependence with nature, exercise the devalued power-to (Allen, 2008; French, 1985).

Every major institution in Western society operates according to these beliefs about power. In business and finance, the aristocratic approach to power reigns with the force of outright military might (Chomsky, 2004) and the more subtle and insidious power of economic policy (Kelly, 2001). Our political system depends on Green's (1998) description of Machiavellian power plays. The power-to of the worker is devalued while the power-over of the CEO is exalted. The productive capacity of workers, of women, and of nature is taken for granted, dismissed, and exploited for the profit of those who wield power-over. Within this paradigm there is no basis with which to value *most* people and *all* of nature. Both become units of production rather than components of an interconnected, sacred whole. What is necessary then is an enhanced valuation of power-to, rather than power-over and the dominance hierarchies thus created.

The Neglected Feminine: Shared Sacredness

Often the term "sacred" is thought to only apply within the context of specific religious traditions. *Shared sacredness*, if viewed from this perspective, is then an oxymoron. How can we share sacredness with those of different faiths? It is even more startling to imagine that we could share sacredness with people of lesser status than ourselves, and with "lower" life forms such as animals, plants, and the non-living natural world, yet this is exactly the ethos we must cultivate if humankind is to survive.

Human existence arises from, and is dependent upon, the natural world (Capra, 1982; Korten, 1999). The matter of both mother and nature, which has been devalued under Aristotelian and Cartesian paradigms, is of increased value when viewed through the characteristically feminine focus on relationships and interdependence.

> Women have permeable boundaries. Perhaps it is the experience of our bodies in touch with the bodies of others that makes it hard for us to close down our psyches. Perhaps it is genetic. Or both. Or something else. But our bodies feel the irrevocable connection of the tides with our cycles of monthly bleeding. And in lovemaking we can be penetrated and receive another. And with pregnancy we carry another for nine full moons, more or less. When we separate from that other, we can feed it from our own body. And later the cycles that tie us to the moon and tides stop. All this is true whether we give birth or not, have sex or not. The possibility is what creates the openness, and the openness is a precious gift.
>
> The distinct flavor of experience which comes with this gift shapes how we perceive reality, how we act, how we create, and what we *value* [emphasis added]. And more than anything else women value relationships. We blend and weave and combine and sustain all kinds of relationships, and this work, this webmaking not only shapes our lives but makes us profoundly vulnerable to the needs of others (Anderson & Hopkins, 1992: 183-184).

We thrive when in connection with others (French, 1986). Without the natural world, without connection to others, we would not exist and we would not thrive. In recognizing this fundamental interdependence among humans—and between humans and nature—we begin to move toward healing and wholeness (Garrison, 2000; French, 1986; Anderson & Hopkins, 1992; Capra, 1982), toward shared sacredness.

A more modern basis for shared sacredness might, ironically, come from mechanistic, reductionist, empirical science. According to Anderson and Hopkins (1992: 178), "Relationship is an incontrovertible fact of life that scientists have been demonstrating repeatedly over the last half century. From the most advanced scholarship in fields from biophysics to botany, archeology to cell science, sociology to neurophysiology, we are learning that we belong to one another and are sustained by each other." Korten (1999) and Capra (1982) argue that an increasing recognition of the interrelatedness of everything in the universe is requisite for cultural transformations that would make continued human existence more likely. This attention to relationship is accomplished in both texts in a similar way to the presentation of Sagan's (1980) series, *Cosmos.*

As carbon-based life forms, our bodies and our minds are made of the dust of stars that became supernovae long ago, making the stars themselves our oldest relatives.[5] All life operates on the principle of emergence that defies entropy, instead forming ever-greater levels of complexity, perpetually evolving, and thereby surrounding us and pervading us with paradox and mystery. We are not unlike the universe itself. We have a lifespan and we will one day cease to exist.[6] When we die, the matter that makes up our bodies will be transformed throughout the biosphere. Eventually some of our atoms may become sedimentary rock. The energy that gives us life will be conserved, radiating throughout the universe indefinitely. Ultimately, our deaths are but a small sacrifice, an inevitable transition necessary for others who are not yet born. It is this connectedness, interdependence, and relationship, with the ancient stars, the non-human world around us, and the distant future, that provides the context for human existence. Within this paradigm, there is little room for dominance. How could we be less than the stars that shine, or greater than the humble soil? Given enough time, each of us is both star and soil.

Shared sacredness cannot, in and of itself, fix every environmental and social problem we face. It is not a panacea, but it is requisite for sustained social transformation which can lead to increased concern for the environment and for the ability of future generations to depend on a *healthy* environment (Korten, 1999; Capra, 1982). It is but one facet of a multifaceted approach to systemic challenges (Gardner & Stern, 2002), but is essential nonetheless. If we are to continue to have a place in this Universe we must begin to see it as sacred, and we must begin to see ourselves as an interdependent part of it. We must seek ethical systems which recognize the sacred in *all* things, rather than separating the sacred out of the physical. We must not place ourselves as separate and above, thereby excusing any exploitation of humanity and nature. We cannot heal ourselves, or our planet, unless we seek wholeness. That wholeness depends on recognizing that we are all connected, all relatives, all sacred. When we do the work to transform our dominant paradigms, from primarily dominance-based toward shared sacredness, we prepare ourselves to act as transformative change agents, ready to contribute to the whole, not as the dominant and separate master, but as the embedded, sacred being, co-creating wholeness, with awe and reverence for all.

The Promise of Reconciliation

Feminine ethics may have arisen from the bodily experiences of females who invariably carry, bear, and nurture the most vulnerable among us (Ruddick, 1989; Anderson & Hopkins, 1992). The feminine ethos arises from these bodily experiences in which the female *surrenders* to the power of nature, her connectedness to, and dependence on it, and her role as the primary nurturer and sustainer of the helpless newborn. This role, as mother, has defined women perhaps longer than the existence of patriarchy (French, 1986; Stone, 1976; Walker, 1983). Through this role, the female is brought face to face with her helplessness in the face of nature, her importance to it, her dependence upon it, and her ability to perpetuate it. She comes to understand her place as an interdependent link in the web of life (French, 1983). She exists within the matrix of nature from which she takes form, and through which she creates new life.

Since the dawn of recorded history, men have constructed meaning and values in their lives by establishing dominance, and engaging in conflict and war. This may have resulted from the male's disconnect from the biological drama of the creation of new life (French, 1983) and from the social role of warrior-protector that men have been conditioned to fulfill (Farrell, 1994). Biologically removed from the suffering, risk, pain, and joys of motherhood, men constructed systems of meaning from the physical activities in which men excelled (French, 1983). This drive to create a sphere, in which males excel and could exert control over others, parallel to the domestic sphere where women as mothers ruled, has resulted in the emergence of increasingly complex civilizations. The narrow focus on empiricism and the cold, uncaring, approach to the natural world thus engendered, has brought us untold technological advances. The ethos of dominance has not been entirely destructive, but has also played a creative role simultaneously. The transformation most necessary today is to capture the most beneficial aspects of dominance and materialism while minimizing their destructive capacities by striking a dynamic balance with the values of interdependence and shared sacredness.

The highest sacrifice the male is thought capable of is the sacrifice of his own life, in war, to protect those who are more vulnerable (Farrell, 1994). While the female biosocial role is to work with nature to create new life, the male's cultural-social role is to act against nature to destroy life, eventually to remove himself from nature entirely as he sacrifices his own life (Farrell, 1994). The male ethos is thus expressed as a presumed position of power-over when the male determines how and when the life, which arises from "Mother Nature," is destroyed at the hands of man. The problem with such a construct is that it only allows men the experience of direct relation to nature in an antagonistic, destructive, and painful way. Man and nature, and man and man, become enemies rather than part of one sacred whole. This antagonism toward the natural world is not only found in the male tendency to wage war, but also in our economic and religious institutions (Capra, 1982; Korten, 1999). In his believed separateness, man has become not creator, nor sustainer, but destroyer. It is this fictional divide, between man and nature, which must be healed. Man must come to see himself as a part of nature, and to see nature as a valuable and adored extension of himself. As Aldous Huxley once stated, "our present economic, social and international arrangements are based, in large measure, upon organized lovelessness," (1944: 93). It is this lovelessness that must be rectified.

One way to rectify such lovelessness might lie in the theory of Trompenaars and Hampden-Turner (2001) who articulate a strategy for reconciling cultural differences, "values in tension, which appear, at first glance, to be negations of one another can, in fact, work in synergy" (p. 4). This process of reconciliation is one part of a three-part process known as the three R's: recognize, respect, and reconcile (Trompenaars & Hampden-Turner, 2001; Trompenaars & Wooliams, 2003). We have completed the first two steps of recognizing that there exists a difference between the ethos of masculinity and femininity, and recognizing that each has a value and purpose. What is missing is permission for both to be embraced by all individuals and institutions. It is this final step that must be accomplished if we are to embrace the beneficial effects of both male and female ethics and ways of being, civilization and nature.

Individuals, both masculine and feminine, have the innate capacity to express both feminine and masculine ethics, attitudes, and actions (Bem, 1974) and it is this ability that must be nurtured.

One must repeatedly remember that nothing is easily labeled and done with. Society has changed, and continues to see positive transformations that free both women and men to live according to their own inner compass. Members of both genders have suffered horribly as members of a dominance hierarchy. Each has been enslaved to the other, and to society, and each longs for freedom to live authentic, connected lives (Farrell, 1994; French, 1985). It is, however, important to recognize that until women are empowered to control their own reproductive lives, we will continue to face the environmental, social, and political challenges of increased population pressures. Only when women exercise control over their fertility can we hope to address many of our most pressing environmental and social challenges (Bandarage, 1997). At the same time men must also be given permission to control their own lives (Farrell, 1994), to choose or reject a militarized and violent masculinity, a masculinity at odds with other humans and with nature (French, 1985). As we move away from strict gender roles, and as we assign greater value to the feminine archetype, our culture may approach wholeness, an ethical ground that is capable of sustaining, rather than destroying, life.

Notes

1. The United Nations Development Programme (Watkins, 2007/2008) cites world population as having increased from around four billion in 1975 to 6.5 billion, thirty years later, in 2005. Al Gore (2006) puts this population increase in an even larger context. For hundreds of thousands of years, from the emergence of the first modern humans in 160,000 BC to AD 1776 there were less than one billion people on the planet. In the short two hundred years from AD 1776 to AD 1976, the human population increased from one billion to over four billion. What had taken millennia now occurred within decades, with startling consequences.
2. The inequalities of the global economic system are evident, but it is beyond the scope of this chapter to elucidate the various workings and effects of said system. The author, in formulating this article, was influenced by several authors: Brown (2001), Domhoff (2006), Henderson (1999), Kelly (2001), McMichael (2008), and Smith (1776/2003), among others.
3. It is beyond the scope of this chapter to list the extensive and dynamic ways in which Earth's biosphere is being compromised. For a comprehensive understanding of the challenges and threats facing life on planet Earth refer to Brown (2003), Rischard (2002), Diamond (2005), and Gore (2006). For resources specifically addressing climate change refer to Gore (2006), and Watkins (2007/2008). Additionally, a comprehensive set of potential solutions are recommended by Steffen (2008), Rischard (2002), Brown (2003), Korten (1999), Kelly (2001), and Gore (2006).

4. For the purposes of this chapter, feminine and masculine ethics and ways of being are defined as the dominant culture generally defines them.
5. All elements heavier than lithium (including carbon–of which we are fundamentally made) were forged in the nuclei of stars during the process of nuclear fusion. The Earth itself was accreted approximately 4.5 billion years ago from interstellar dust that had been created by stars that had exploded into supernovae sometime in the ancient past (Sagan, 1980).
6. While the ultimate fate of the universe is not known, of the two potential futures, the Big Crunch and endless expansion, current observations support the latter (Hinshaw, 2008).

References

Adams, C. J. (1993) *Ecofeminism and the sacred.* New York: Continuum.

Allen, A. "Feminist Perspectives on Power", *The Stanford Encyclopedia of Philosophy (Fall 2008 Edition)*, Edward N. Zalta (ed.), URL = <http://plato.stanford.edu/archives/fall2008/entries/feminist-power/>.

Anderson, S., and Hopkins, P. (1992). *The feminine face of god: The unfolding of the sacred in women.* New York; Bantam Books.

Aptheker, B. (2003). Introduction to Feminisms (course video), retrieved on April 12, 2009, from http://www.introtofem.org/.

Aristotle. *Generation of animals,* edited and translated by A. L. Peck. London, 1953 (Loeb, Classical Library) PA3611 .A73 1953.

Bakan, D. (1966). *The duality of human existence.* Chicago: Rand McNally.

Bandarage, A. (1997). *Women, population and global crisis: A political-economic analysis.* London: Zed Books.

Bem, S. (1974). "The measurement of psychological androgyny." *Journal of Consulting and Clinical Psychology, 42,* 155-162.

Brown, L. R. (2001). *Eco-economy: Building an economy for the earth.* New York: W.W. Norton & Company.

Brown, L. R. (2003). *Plan B: Rescuing a planet under stress and a civilization in trouble.* New York: Norton and Company.

Capra, F. (1982). *The turning point: Science, society, and the rising culture.* New York: Bantam.

Chomsky, N. (2004). *Hegemony or survival: America's quest for global dominance (American empire project)* New York: Henry Holt and Company.

Diamnod, J. (2005). *Collapse: How societies choose to fail or succeed.* New York: Penguin Books.

Domhoff, G. W. (2006). *Who rules America?: Power, politics, & social change.* San Francisco, CA: McGraw Hill.

Farrell, W. (1994). *The myth of male power.* New York: Berkeley.

French, M. (1986). *Beyond power: On women, men, and morals.* New York: Ballantine Books.

Gardner, G. (2002). *Invoking the spirit: Religion and spirituality in the quest for a sustainable world.* Danvers, MA: World Watch Institute.

Gardner, G., and Stern, P. (2002). *Environmental problems and human behavior (2nd ed).* Boston: Pearson Custom.

Garrison, J. (2000). *Civilization and the transformation of power.* New York: Paraview.

Goldberg, S. (1993). *Why men rule: A theory of male dominance.* Peru, IL: Open Court.

Gore, A. (2006). *An inconvenient truth: The planetary emergency of global warming and what we can do about it.* New York: Rodale.
Green, R. (1998). *The 48 Laws of Power.* New York: Penguin.
Harper, D. (2001). "Matter"—definition, *Online Etymology Dictionary.* Accessed April 12, 2009, from http://www.etymonline.com/index.php?term=matter.
Hausman, G. (1993). *Tunkashila: From the borth of turtle island to the blood of wounded knee.* New York: St. Martin's Press.
Henderson, H. (1999). *Beyond globalization: Shaping a sustainable economy.* West Hartford, CT: Kumarian Press Inc.
Heyd, T. (2007). *Encountering nature: Toward an environmental culture.* Hampshire, England: Ashgate.
Hinshaw, G. (2008). *What is the ultimate fate of the universe?* NASA. Retrieved on April 13, from http://map.gsfc.nasa.gov/universe/uni_fate.html.
Hite, S. (1994). *The Hite report on the family.* New York: Grove Press.
Hofstede, G. (2001). *Culture's consequences: Comparing values, behaviors, institutions and organizations across nations.* London: Sage.
Hofstede, G. (ed.) (1998). *Masculinity and femininity: The taboo dimension of national cultures.* Thousand Oaks, CA: Sage.
Huxley, A. (1944). *The perennial philosophy.* New York: Harper and Brothers.
Kelly, M. (2001). *The divine right of capital: Dethroning the corporate aristocracy.* San Francisco, CA: Berrett Koehler Publishers.
Korten, D. C. (1999). *The post corporate world.* San Francisco, CA: Berrett Koehler Publishers.
Lerner, G. (1986). *Patriarchy and the control of womanhood: The creation of patriarchy.* Oxford: Oxford University Press.
Machiavelli, N. (1513). *The prince.* Introduction by Thompson, N. H. (1992). Mineola, NY: Courier Dover Publications.
Mander, J. & Tauli-Corpuz, V. (2006). *Paradigm wars: Indigenous people's resistance to globalization.* Berkeley, CA: University of California Press.
Milhausen, R., and Herold, E. (1999). "Does the Sexual Double Standard Still Exist? Perceptions of University Women." *Journal of Sex Research.* Accessed April 14, 2009, from http://findarticles.com/p/articles/mi_m2372/is_4_36/ai_58459537/.
Mills, W. (1956). *The power elite.* Oxford: Oxford Press.
Mir-Hosseini, Z. (1999). *Islam and gender: The religious debate in contemporary Iran.* Princeton, NJ: Princeton University Press.
Merriam Webster Dictionary http://www.merriam-webster.com/dictionary/dominant.
McMichael, P. (2008). *Development and social change: A global perspective.* Los Angeles: Pine Forge Press.
Nasr, S. H. (1989). *Knowledge and the sacred.* Albany, NY: State University of New York Press.
Pearson, c. (2008, May 7). *Go all the way upstream...* Retrieved August 27, 2008, from Real Wealth: We Have a Choice. http://www.realwealtheconomy.com/?paged=2.
Ray P. & Anderson S. R. (2000). *The cultural creatives: How 50 million people are changing the world.* New York: Three Rivers Press.
Rischard, J. F. (2002). *High Noon: 20 global problems, 20 years to solve them.* New York: Basic Books.
Ruether, R. (1993). "Ecofeminism: Symbolic and Social Connections of the Oppression of Women and the Domination of Nature," in Adams, C. (ed.) *Ecofeminism and the Sacred.* New York: Continuum.
Ruddick, S. (1989). *Maternal Thinking: Toward a politics of peace.* Boston: Beacon.

Sagan, C. (1980). *Cosmos.* New York: Ballantine.
Smith, A. (1776/2003). *The wealth of nations.* New York: Bantam Dell.
Smith, H. (1991). *The world's religions.* San Francisco: Harper San Francisco.
Smith, L. T. (1999). *Decolonizing methodologies: Research and indigenous peoples.* London: Zed Books Ltd.
Steffen, A. (2008). *World changing: A user's guide for the 21st century.* New York: Abrams.
Stone, M. (1976). *When God was a woman.* Fort Washington, PA: Harvest Book Company.
Trompenaars, F., and Hampden-Turner, C. (2001). *21 leaders for the 21st century: How innovative leaders manage in the digital age.* Oxford, UK: Capstone.
Trompenaars, F., and Wooliams, P. (2003). *Business across cultures.* West Sussex, UK: Capstone.
U. S. Department of State. (n.d.). *The Truman Doctrine 1947.* Retrieved September 21, 2008, from U.S. Department of State: http://www.state.gov/r/pa/ho/time/cwr/82210.htm.
Walker, B. (1983). *The woman's encyclopedia of myths and secrets.* San Francisco: Harper.
Watkins, K. (2007/2008). *Human development report.* New York: United Nations Development Programme (UNDP) and Palgrave Macmillan.

Exploring Ethical and Religious Attitudes to Sustainability

Robin Attfield

Defining Sustainability

First, a few words are in place about what is meant by "sustainability." Here in Wales, the word conjures up images of limiting carbon emissions through means such as loft insulation or the use of sustainable sources of energy. People would in any case have in mind processes or systems that are benign and conducive to an enhanced quality of life, certainly for human beings and possibly for nonhuman creatures or for ecosystems. But not everything sustainable is desirable. For example, slavery was arguably a sustainable system of production, but highly oppressive. This being so, we need to come clean about which values are being pursued in the name of sustainability. Otherwise there might be no ethical or religious case for sustainability at all, or there might even be a religious or quasi-ethical case for kinds of sustainability of an undesirable kind, such as sustainable kinds of authoritarianism.

It might be replied that environmental sustainability is free of problems, and beyond objection. But there was environmental sustainability in ancient Egypt, where the annual flooding of the Nile enabled the population to support itself. This, however, was accomplished for many centuries through the labor of slaves, and was therefore an objectionable system. Someone might protest at this point that no one in the modern world objects to environmental sustainability on any such grounds; but they would be wrong. For the Indian philosopher Ramachandra Guha (1989) has objected to the application of Western environmentalism to Third World countries in forms that disregard the needs of local people. More generally, sustainability is unlikely to be defensible unless it is coupled with fairness, and unless its proponents have regard to the basic needs of affected parties, whether human or nonhuman.

This in turn helps explain the need for policies not so much of sustainability as of sustainable development. The phrase "sustainable development" was the theme of the Brundtland Report of 1987, and was endorsed by most countries in the Rio Conference on Environment and Development of 1992. "Development" is tacitly short for social, or economic, development, the kind of development in which human needs are met, or in which changes are made in that direction; it can also describe the state of affairs in which these needs are already realized. The inclusion of the term "sustainable" involved, or was originally intended to involve, the requirement that such development should be sustainable not only economically but also socially and environmentally; in other words, development was to be durable, and to be combined with environmental sustainability. This joint objective was to recognize not only the value of satisfying basic human needs, but also those of nonhuman species, including needs for viable habitats. (If nonhuman needs were not included, then this objective would still be open to the charge of treating human beings as having exclusive value and importance, or of anthropocentrism; but there is every reason not to lay ourselves open to this criticism.)

Intergenerational Equity

So if we use "sustainability" as a codeword for "sustainable development," we are much closer to something for which there is a religious or an ethical case. However, before we can come to any such case or cases, we need to consider Wilfred Beckerman's (1999) objection that sustainable development is radically unfair, in particular to earlier rather than later generations. For it surely implies that there should be continual improvement in quality of life, and this would mean that you fare better the later you live, and the less well the earlier you live. And if this is what it means, then there is surely no obligation, religious or ethical, to promote it. However, Beckerman's interpretation is a travesty of what was meant in the Brundtland Report. A corrective has been supplied by Alan Carter (2000). "Sustainable development" should be taken to involve processes of moving away from poverty and underdevelopment which are prerequisites of sustainability, and lead to a sustainable condition of development (the state in which basic needs are satisfied) which is sustainable in the sense that it can then be maintained indefinitely, economically, socially and environmentally. In this way, the objection about unfairness is overcome, and sustainable development can now be advocated in Carter's sense. We can even use "sustainability" to mean the same thing, as long as we do not forget the vital qualifications about development and about equity, which are needed, as has been seen, to avoid "sustainability" being open to objection in many of its forms.

Ethical Approaches to Sustainability

At last, we can turn to the ethical and religious case for such sustainability. Here I say "case," because that is the kind of thing (grounds, reasons and arguments) that philosophers study. If the question were about the ethical stance of (say) most people now alive in Wales or in Britain, then it would require a sociological answer, and one that I am not qualified to give. But if it concerns the resources of ethical traditions, then I can at least make a start.

There are three ethical traditions that would struggle to give full support to such sustainability. First there is the one that bases ethics on a contract, whether an historical contract or a hypothetical one. The general idea is that what it is right to do is what would be agreed in some fair bargaining situation. John Rawls (1971) was the most prominent twentieth-century advocate of this approach. One problem is that nonhuman animals are not capable of entering into agreements, and are excluded. Another problem, and one that Rawls strove to overcome in his book *A Theory of Justice*, is that future generations will only be included if the people who form the agreement are inclined to represent them, and Rawls did not initially want these people to look out for any interests except their own. (After all, if self-interested people would agree to run society by certain rules, then it seems reasonable to regard them as fair rules.) Rawls tried to get over this problem by modifying the so-called 'initial position' in which the original agreement takes place, but never really managed to do so. Yet sustainability will not be adequately supported unless the interests of every generation, including future generations, count. Not everyone accepts these criticisms of contract-theory, and there are many influential Rawlsians still around. Yet the two problems mentioned, provision for non-human animals and provision for future generations, suggest that no satisfactory basis for sustainability is here to be found.

Another tradition attempts to base all ethical considerations on rights, and the responsibilities that correspond to them. Some of these theorists restrict rights to human beings, and for these the problem of provision for nonhumans applies just as much as it did for contractarians. But there is also a problem with future generations, not only for rights-theorists who are anthropocentric but even for the ones that are not. To have a right, someone or something must be capable of being treated according to their rights or not being so treated. But, generally speaking, most future people would exist only in one possible future, for if current agents (prospective parents, for example) conceive and give birth from different eggs or sperms, then different people and a different future come about. (The mere timing of conception is enough to make the difference between one possible person and another being generated.) Since having rights involves existing in more than one future, but most possible people exist in one future only, most possible people of the future would seem not to have rights. Admittedly they will have rights if they come into existence, on just the same basis as that on which we have rights, for by then it will be possible for them to have their interests satisfied or neglected by their contemporaries. But if the case for taking their interests into account in advance of their lives depends on their rights, then it is a case that collapses. Yet once again, sustainability will not be adequately supported unless the interests of every generation, including future generations, count. From this it may be concluded that ethics must rest on something beyond rights; for it is possible to take seriously the interests of coming generations, but this is not possible on an approach that bases everything on rights.

Yet a third tradition has problems with sustainability. What I have in mind here is Kantianism. Kant's Categorical Imperative (1948) bids us behave only in a way in which we could rationally desire every agent to behave. So far, so good, at least where the environment is concerned; some Kantians have taken this to imply that if certain behavior, such as emitting carbon dioxide unnecessarily, would be undesirable if done by everyone, then it should not be done by anyone. Yet this is a strange principle, for there are many deeds which would be disastrous if done by everyone (like becoming a university student, for example), but which are unobjectionable if done by one or even by many individuals. Besides, when Kant rephrases his Categorical Imperative, he makes it say that we should never treat humanity as a means only, but always as an end. This would be admirable if it were just a sectoral principle, but as a foundational principle it unfortunately excludes any concern for nonhuman creatures. Again, it is less than clear how the interests of future generations are to be safeguarded on this basis, for we cannot treat them as individuals prior to their existence, and, if the reasoning advanced in connection with rights is correct, then we cannot treat unborn individuals better or worse at all, as opposed to determining which ones will live and how well one population or another will fare. Therefore Kantianism, which initially seemed so promising, cannot really uphold sustainability.

A fourth tradition appears to fare better. The philosophy of Aristotle (1953) places the virtues at the center of the moral life, and claims that the criterion of rightness is what the virtuous person would do. So if the virtuous person leads a sensitive, low-consumption life, consonant with sustainability, that is what it is right to do. Aristotle's own list of virtues in *Nicomachaean Ethics* does not quite correspond to these, but his modern supporters (advocates of "virtue ethics") sometimes revise the list of virtues so as to enthrone green ones. But unfortunately there is surely no guarantee that what the virtuous person does would be right. So complex is our technological world that many virtuous people would make sincere but disastrous mistakes, from which their virtue would entirely fail to immunize them. Besides, we surely need some criterion of what the virtues are which is independent of what the virtuous person would do, a criterion such as the impacts of their actions. Hence, while the virtues remain important, and possibly central, and while what Aristotle says about them is largely sound, we need some other basis for discovering or deciding what it is right or justified to do.

The approach that makes rightness depend on impacts is consequentialism. The best-known version of consequentialism is utilitarianism, which seeks to maximize the balance of happiness over misery, and to this there are numerous well-known objections. But consequentialism can instead be targeted at maximizing well-being or quality of life (broader concepts both than happiness), or, where these are central examples of intrinsically valuable states of affairs, at maximizing intrinsic value. Intrinsic value can also be found in the flourishing of nonhuman creatures, and of creatures of the future as well as of the present, and maximizing this obviously involves preserving the ecosystems on which they depend. In this inclusive, biocentric form, consequentialism already looks promising from an environmentalist perspective.

It would, of course, be prohibitively difficult to be expected to calculate the impacts of all one's actions. But here consequentialists can appeal to the impacts not of individual's acts, but of whole practices (practices such as recycling), and make actions right that comply with the practices with the best overall impacts. In this way, consequentialism secures some of the advantages that seem to belong to Kantianism, with its universal principles, and to rights theory, which in effect demands across-the-board abstinence from practices such as torture; as long as principles or abstentions have better net impacts than alternative principles or practices, consequentialism can uphold them, and supply a satisfactory reason for doing so as well, as also for any exception clauses required to make the practice as beneficial as possible. Practice-consequentialism avoids many of the pitfalls of act-utilitarianism, and biocentric practice-consequentialism has a broad enough value-theory to avoid most if not all of the others. For example, it supplies a possible basis for testing the desirability of legislation and regulation, including environmental legislation and regulation. Contractarianism and Kantianism might be thought to show more respect for human autonomy, but there is nothing within consequentialism to prevent recognition of the intrinsic value of such autonomy, and thus the rejection of practices that would be unduly paternalistic. However, paternalism is not the same as interventionism, and the latter would be likely to be supported, at least wherever ecological thresholds were in danger of being crossed or species in danger of being eradicated.

What is still missing here is recognition of the virtues. Aristotle (1953) was right in holding that there are certain virtues such as courage that every society will need. Indeed certain other virtues will also be needed both by individuals, corporations and governments if ecological challenges are going to be met, for all these agents are going to need dispositions to behave in a manner that takes account of relevant characteristic impacts. No longer can it be claimed that behavior is right simply on the basis that this is what the virtuous person would do; we need an account of why certain virtues are virtuous, and thus why behavior of this kind is right. For example, the virtues needed are going to include green virtues, recognized as such because habitual compliance with them helps preserve species and habitats while minimizing the environmental footprint of the agent concerned. What is needed here is a further broadening of practice-consequentialism so as to embody virtue-consequentialism, as recently advocated in an article by Dale Jamieson (2007). Needless to say, this should not be anthropocentric virtue-consequentialism, even though this is the variety that philosophers such as Aristotle and Kant might well, at a pinch, have preferred (if they could have overcome their antagonism to consequentialism in the first place); rather it should be biocentric consequentialism, as explained earlier. This is the kind of ethic that in my view will increasingly be needed, and which, even if it is not given explicit recognition, is likely to receive increasing implicit support.

Theistic Religious Approaches to Sustainability

However, I was asked to comment on religious approaches as well as ethical ones. There is time here to comment only on theistic religions (such as Judaism, Christianity, and Islam), rather than other religions including eastern religions. Such religions have been accused of being anthropocentric, but this charge does not stand up. The Old Testament is common ground to all three of the theistic religions, and it is impossible to read passages such as Psalm 104 as upholding an anthropocentric approach. Nor do these religions suggest that humanity may treat planetary nature as it pleases. They all recognize that humanity is in a position of authority over planetary nature, but they all equally treat this as responsibility before God, holding that the earth belongs to God and not to us. This responsibility is to be exercised ethically, that is, sensitively, justly and taking into account coming as well as current generations, and species other than our own. These themes have received new emphasis in recent decades with the rediscovery of the stance called "stewardship."

All kinds of objections have been presented to stewardship. For example, it is claimed that it cannot extend to stars and galaxies other than our own. Yet it needs to extend to the entire zone that humanity can influence, since otherwise there would be human power without responsibility. There again, it is objected that the role of humanity should often be one of letting-be rather than stewardship. But this assumes that stewardship cannot justify inaction rather than constant intervention. Where ecosystems are best preserved by letting-be, stewardship will go along with this policy. To consider one last objection, it is sometimes held that stewardship involves managerialism and an instrumental attitude to nature. But this is to misread the role of a steward, which can also be expressed as that of a trustee. The valuable creatures that are in the charge of human stewards need not be treated as dispensable resources, and can be recognized as the treasures and the bearers of intrinsic value that they often are.

In short, there is nothing to prevent the religious ethic of stewardship being harnessed to the secular ethic of biocentric consequentialism, a consequentialism that can also be one of beneficent practices and virtues. When stewardship is interpreted in this way, non-religious people can, of course, endorse it. I have long been arguing that secular versions of stewardship supply a comprehensible framework for human self-understanding; this approach, after all, embodies the virtue of humility towards nature, which may itself be one of the beneficent virtues just mentioned.

(See Attfield, 2003). Nevertheless, in my own view, a religiously informed self-understanding such as stewardship supplies a more coherent overall approach to nature and to life, as well as to sustainability, with a greater motivational capacity than its secular counterpart, which may well make it a more effective system of belief for promoting sustainable behavior.

References

Aristotle. (1953). *Nichomachean Ethics*, trans. J.A.K. Thomson as *The Ethics of Aristotle*, Harmondsworth: Penguin.

Attfield, R. (1995). *Value, Obligation and Meta-Ethics*, Amsterdam and Atlanta, GA: Éditions Rodopi.

Attfield, R. (1999). *The Ethics of the Global Environment*, Edinburgh: Edinburgh University Press and West Lafayette, IN: Purdue University Press.

Attfield, R. (2003). *Environmental Ethics*, Cambridge, UK and Malden, MA: Polity.

Beckerman, W. (1999). "Sustainable Development and Our Obligations to Future Generations'," in Andrew Dobson (ed.), *Fairness and Futurity*, Oxford: Oxford University Press, 71-117.

Carter, A. (2000). "Distributive Justice and Environmental Sustainability," *The Heythrop Journal*, XLI, 449-460.

Guha, R. (1989). "Radical American Environmentalism and Wilderness Preservation: A Third World Critique" *Environmental Ethics*, 11, 1989, 71-83.

Jamieson, D. (2007). "When Utilitarians Should Be Virtue Theorists," *Utilitas*, 19, 160-183.

Kant, I. (1948). *Groundwork of the Metaphysic of Morals*, trans. H.J. Paton as *The Moral Law*, London: Hutchinson University Library.

Rawls, J., *A Theory of Justice*, Cambridge, MA: Harvard University Press, 1971

World Commission on Environment and Development. (1987). *Our Common Future*, Oxford: Oxford University Press.

Against So Much Money and Power, Can the Sustainability Movement Succeed?

Ellen Gaddy

The Centralized Monetary System as an Impediment to Sustainability

In "Against So Much Money and Power, Can the Peace Movement Succeed?" Marc Pilisuk and I (in press) outlined the challenges of creating a peaceful and sustainable worldview within the existing monetary, economic, and political system—a system that is dominated by overlapping hierarchical networks of power between finance, government, industry, and the military. Our rationale was informed by the wisdom of Audre Lorde, who once pointed out that it is impossible to dismantle the master's house using the master's tools. Lorde believed if a movement's *only* approach for social transformation is embedded in the existing system, the efficacy of the movement would be limited; limited because social change within the existing system would come about slowly and with great concessions, and limited because the master's hierarchical structuring of society would remain intact.

The core of our thesis pertained to the master's most effective tool for maintaining its hierarchical power structure: the centralized monetary system. The centralized monetary system operates both nationally and internationally and maintains a monopoly over the money supply. Those that are deemed uncreditworthy, such as the poor in both developed and underdeveloped countries, are largely excluded from participation in this system. Those that are allowed to participate are shackled to the system in the form of interest bearing debt and punished with bankruptcy if they cannot keep apace of their payments.

Money, in the centralized monetary system, is said to be politicized for two reasons. One, the central bank is the bank of the state and provides funding for all the state's endeavors, including military endeavors. And, two, a "revolving door" exists between government and finance to ensure that this system remains intact. We exemplified this revolving door by outlining the prevalence of notable former Goldman Sachs executives in influential government positions. Goldman Sachs is the world's most powerful investment bank, and its former executives were instrumental in designing the multi-trillion dollar bailout, which, one could not help but notice, benefited Goldman Sachs and friends over other financial institutions and the American people. For generations to come, the American people must foot the bill for Wall Street's recklessness in the form of taxes and inflation. And so, the politicization of money greatly limits the degree to which monetary change can be brought about within the existing system.

Given that the centralized monetary system is built upon interest-bearing exponentially increasing debt, requiring a growth imperative for repayment, credit is disproportionately extended to those short-term projects, ventures, and investment schemes that promise the highest rates of return. Unfortunately, short-term, high-yield investments are usually speculative in nature, and unbridled speculation combined with an inflationary monetary system results in inevitable cycles of boom and bust. During bust cycles, the financially unsophisticated and vulnerable see their wealth and resources disappear, only to reappear in the bank accounts and holdings of those "too big to fail" financial institutions who are savvy at manipulating economic bust cycles in a way that minimizes their losses and maximizes their gains. Meanwhile, credit dries up, which sets off a domino effect of defaults, foreclosures, and bankruptcies. In this way, the centralized monetary system guarantees the redistribution of wealth from the poor and middle class to the rich.

However, an even greater problem of the centralized monetary system is that its built-in growth imperative demands exponential economic growth in order to sustain itself. This built-in growth imperative results in the rampant consumption of resources and the destruction of the ecological environment. So, not only does the centralized monetary system consolidate wealth in hierarchies of power, it also directly opposes the creation of a sustainable worldview. The future success of the sustainability movement depends on its ability to address the problem of the monopolized, politicized, and centralized monetary system.

The Sustainable Crossroads

The Summer 2009 issue of *Yes! Magazine*, titled "The New Economy Starts Here: Why This Crisis May Be Our Best Chance," demonstrated the hope, shared by most members of the sustainability movement, that the economic crisis of 2008-2009, for all its hardship and heartache, would provide the impetus required to bring about sustainability's "tipping point" (Gladwell, 2002). Instead, the U.S. government bailed out Wall Street without the imposition of increased oversight and regulation, the stock market made a historically rapid recovery, and then, a few months later, the healthcare debate stole the political limelight and succeeded in sidelining calls for serious economic and monetary reform. And so, as the panic of the last year and a half recedes from memory and society slips back into the pre-crisis status quo, so too does the opportunity for actualizing sustainability's tipping point. As the first decade of the new millennium comes to a close, we find ourselves only slightly closer to the creation of a sustainable world.

That the economic crisis did not turn out to be "our best chance" should have been expected. It reflects what we know about human behavior under stressful conditions. Psychological evidence abounds to suggest that periods of chaos and crisis are just as likely, if not more than likely, to stimulate an unmanageable amount of anxiety and terror which result in an *increased defense* of the current worldview and its inherent values, beliefs, and concomitant behavior rather that an *increased acceptance* of a new worldview with its new values, beliefs, and concomitant behavior (Pyszczynski, Solomon, & Greenberg, 2003). Also, history suggests that ruling elites know how to utilize these periods of collective shock and crisis to strengthen their position of influence and dominance within the current system (Klein, 2007). Those at the top of financial, corporate, and government hierarchies dominate the master's game and are not likely to make concessions over a manageable economic crisis.

Though the economic crisis did not turn out to be our best chance, it nonetheless served to alert the sustainability movement of three critical shortcomings. First, the sustainability movement is limited by strategies and approaches whose implementation overwhelmingly rely on changes within the existing system, such as nationalizing the money supply or ending the Federal Reserve. Second, these embedded approaches contain no means of propagating and reinforcing the sustainable mindset required to create a peaceful and sustainable world. And, third, while Paul Hawken's *Blessed Unrest* was optimistic in regard to the efficacy of the sustainability movement's nebulous, unorchestrated revolution with no leader, no doctrine, and no strategy, one is not readily convinced that this approach will have an effect sufficient to challenge the seemingly omnipotent status quo (Pilisuk & Gaddy, in press). The sophisticated nature of the master's game requires a sophisticated and strategic plan that acknowledges the master's sagaciousness and knows how best to work with it and around it.

Thus, the sustainability movement is at a crossroads. This crossroads is symbolic of the shift from first wave to second wave sustainability. Should the movement continue to focus its energy predominately on unorchestrated approaches that are embedded within the master's system, as is seemingly the philosophy of first wave sustainability? Should the movement branch out and rally support for another monetary system, operated by and for "we the people?" Or should the movement devise and implement a strategy that promotes change both within and outside the existing monetary system? Lewis D. Solomon, a critic of the centralized monetary system, captured the essence of the sustainable crossroads with the following comment: "It is easy enough to envision an ideal society; many, indeed, have done that. But one must deal with the world as it is, rather than as one would like it to be, and this adds immeasurably to the problem" (1996: 1). The purpose of this chapter is to provide answers to these questions and aid the process of constructing a strategy for second wave sustainability that heeds the advice of Audre Lorde and Lewis D. Solomon. But first, a refresher course on money.

The Collective Money Agreement

Distilled to its purest definition, money is merely a community's (global or local) agreement to use something as a means of payment (Lietaer, 2001). All that has been built upon this agreement, central banks, the Federal Reserve, Wall Street, the International Monetary Fund, and the World Bank, encourage two misconceptions; that these institutions are required to create and oversee money, and that the global centralized monetary system is the only way to provide a means of exchange for the global population. Despite the overlay of these imposing hierarchical financial authorities, the most important thing to grasp is that money is whatever a community decides it to be.

Money can, and has, taken many forms such as cowrie shells, pieces of pottery, precious metals, amber, grain, ivory, and cattle (Lietaer, 2000). However, with the expansion of global trade, money in the form of commodities became a cumbersome means of foreign exchange. Eventually, a gold-based monetary system was established which utilized gold coins as money (Zarlenga, 2002). Given its heavy weight, gold also became a cumbersome means of payment, which led to the issue of paper money that was backed by gold. The gold standard ended in the United States during the Great Depression. It was reinstated on a limited basis after World War II and prevailed until 1971, when President Nixon took the U.S. dollar (the global currency) off the gold standard internationally. Since this time, the global monetary system has operated under a fiat standard. Under the fiat standard, money is backed by nothing and is created by a central authority (Lietaer, 2001). Thus, fiat money is the current, collective money agreement.

In our current money agreement, money is characterized by four main features: It is attached to a nation-state or a group of nation-states, it is created out of nothing (fiat money), the majority of money currently in the monetary system began as a bank loan (debt), and all bank loans must be paid back with interest (Lietaer, 2001). Thus, the current collective agreement about money is that, although money is backed by nothing, it requires a central banking authority (in the United States this authority is the Federal Reserve) to oversee its creation and maintain an optimum volume of money in the system, a service that must be compensated for in the form of interest payments. As one proponent of monetary transformation put it: "Under this regime, the people's own credit is privatized and "loaned" back to them at interest" (Greco, 2009: 100).

To add insult to injury, commercial banks, under the authority of the Federal Reserve, are allowed to charge customers a rate of interest that is much higher than the federal funds rate, which is the rate at which commercial banks are allowed to borrow money from the Federal Reserve. On top of this, commercial banks are allowed, via the fractional reserve system, to maintain only a small fraction of their deposits on reserve, while loaning out the rest at interest. And, thanks to the money multiplier, by the time an initial deposit works its way through the banking system as a succession of loans and deposits, an amount of money equal to ten times the amount of the initial deposit will be added to the money supply as a whole, all of it interest-bearing. Thus, the fractional reserve system and the charging of interest greatly increases a commercial bank's bottom line and results in the continuous redistribution of wealth in favor of the monied class (Pilisuk & Gaddy, in press).

Furthermore, when a debtor takes out a loan, an amount of money equal to the principal amount of the loan is added to the monetary system. The amount of money equal to the amount of interest due on the loan is *not* added to the monetary system. In this way, money is kept artificially scarce in the system. As Bernard Lietaer, former senior Central Bank executive and proponent of sustainable monetary systems saw it:

> To put it simply, when you pay back interest on your loan, you are using up someone else's principal. In other words, the device used to create the scarcity indispensable for a bank-debt system to function involves having people compete for the money that has not been created, and penalizes them with bankruptcy whenever they do not succeed (2001: 52).

Due to the artificial scarcity of the money supply, interest directly encourages ecologically unsustainable individual, collective, and institutional behaviors such as competitiveness; hoarding; the accumulation and concentration of wealth, resources, and power; and by favoring short-term speculative investments over long-term sustainable investments (Lietaer, 2001).

In summation, the current money agreement in which a monopolized, politicized, and centralized authority is allowed to create money from nothing, in the form of bank debt, all of it interest bearing, within the fractional reserve banking system, is not conducive for equitable and sustainable economic development that honors the limitations of the ecological environment. Nor is it conducive for propagating and reinforcing sustainable behavior and, by extension, the sustainable mindset required to actualize sustainability's tipping point. The questions, then, are what are the alternatives to the centralized monetary system, and are these alternatives capable of providing a fair, equitable, and sustainable means of exchange for all members of the global population?

Approaches to Sustainable Monetary Systems

Thoughtful individuals have been writing about the problem of the centralized monetary system for decades (see Greco, 1990, 1994, 2001; Lietaer, 2000, 2001; and Solomon, 1996). However, the recent economic crisis effectively increased awareness, curiosity, and knowledge concerning the inherent problems of our global monetary system and resulted in a renewed support of monetary reform and/or alternative monetary systems (see Greco, 2009; Korten, 2009; Paul, 2009; and Pilisuk & Gaddy, in press). The problem, however, is that the sustainability movement has not yet come together and explored this issue thoroughly. While many proposals have been presented, an open line of communication and official forum for debate between proponents of each system has yet to be established. This chapter calls upon the sustainable community to come together and thoroughly examine and debate the different proposals for monetary reformation and transformation, so that the movement, as a whole, might come to a consensus about how best to approach the creation of a monetary system that is equitable, ethical, and sustainable for all people. Following is an outline of the most prominent approaches to creating a sustainable and equitable monetary system.

Nationalize the Money Supply and Issue Debt-Free Money

David Korten is the leading voice within the sustainability movement calling for the nationalization of the money supply. In his most recent book *Agenda For a New Economy: From Phantom Wealth to Real Wealth*, which was inspired by the economic crisis, Korten proposed that the U.S. government nationalize the money supply and issue debt-free money. Money is noninflationary when created in direct proportion to the amount of goods and services entering the market place (Greco, 2009). Thus, instead of *lending* debt money into the system as is the current practice, Korten believes that debt-free credit money should be *spent* into the system in order to pay for public goods and services such as infrastructure, education, and investments in green technology and clean energy. Korten called for a combination of the separation of powers and transparency to govern the integrity of the nationalized system. As Korten viewed it: "An independent federal agency would decide how much new money to create. The Congress would decide how it would be spent. And by law, the whole process would be transparent and open to public scrutiny" (2009: 143). Korten firmly believes that the government issuance of debt-free money is an effective way to reduce taxes, provide public services, minimize inflation, eliminate economic boom and bust cycles, and provide the foundation for a peaceful and sustainable worldview.

Limitations. Korten's first wave solution is not a private, free-market approach to the creation of a sustainable monetary system, which means that it requires a political majority for its implementation However, the degree of politicization of the centralized monetary system suggests that nationalization is an impracticable approach. As Thomas H. Greco, Jr., a leading authority on free-market approaches to monetary and financial transformation explained, "Given the fact that national politics has come so completely under the control of the elite interests, and given the arrangement that has aligned the political interests with the global financial interests, the political approach would seem to have little chance of solving anything" (2009: 110).

Another limitation of nationalization, according to Greco, is that "It fails to recognize the insidious nature of elite power, and that the fundamental problem with the present political money system is the monopolization of credit (money) per se, and not who happens to be the owner of that monopoly" (2009: 110). So long as the money supply remains monopolized by one party or another, so too remains the opportunity for abuse and manipulation of the system.

Thus, Greco argued, the sustainability movement should be wary of expending undue energy and effort on an approach that is both improbable and problematic. Instead, "we should choose to apply our resources to actions that promise the biggest payoff and have the potential to effect timely change" (Greco, 2009:111).

End the Federal Reserve and Reinstitute the Gold Standard

Congressman Ron Paul (2009) is well known for his proposal to end the centralized monetary system in the U.S. by abolishing the Federal Reserve and reinstating the gold standard. Paul is a Libertarian, and as such, he believes that the monetary system should be regulated by the market system. As Paul stated of our current money agreement and the recent multi-trillion bailout, "Just so we are clear: the modern system of money and banking is not a free-market system. It is a system that is half socialized—propped up by government—and one that could never be sustained as it is in a clean market environment" (2009: 18). Like many monetary reformists, Paul is concerned that future economic crises could lead to increased centralization, rather than decentralization, via a worldwide central bank that would oversee and control a global fiat currency. This increased centralization of money and power would further undermine personal and collective liberty and threaten any chance for bottom-up sustainable economic development.

Congressman Paul has been a prominent voice in politics calling for a return to the gold standard, beginning in the late 1970s when he coauthored, with Jesse Helms, the legislation that passed to form the Gold Commission. The purpose of reinstating the gold standard, according to Paul, is to put an end to inflationary fiat money. Pegging a currency to a commodity effectively grounds the currency back into the natural world. According to Paul, this would cap the money supply's elasticity and hinder the rampant currency speculation that results from the global fiat money agreement. Though a firm proponent of gold, Paul is also open to pegging the dollar to a basket of commodities, so long as this initiative arises from the private sector.

By abolishing the Federal Reserve, the centralized money monopoly in the U.S. would come to an end. Paul's post-Fed vision includes multiple currencies and sounder investment opportunities. In this scenario, the market system would regulate the financial industry and the U.S. government would not be allowed to interfere by propping up those speculative and irresponsible financial entities that are considered to be "too big to fail." As Paul stated, "When we unplug the Fed, the dollar will stop its long depreciating trend, international currency values will stop fluctuating wildly, banking will no longer be a dice game, and financial power will cease to gravitate toward a small circle of government-connected insiders" (2009: 202).

Paul realizes that it would be difficult to unplug the Fed all at once. Thus, he recommends following these small and manageable steps towards sound money: 1) Put a cap on the elasticity of the money supply; 2) restrict the Fed's ability to buy U.S. Treasury bonds on the open-market; 3) deny the Fed the ability to monetize debt; 4) disallow the Fed from participating in economic planning; 5) disallow the Fed to bail out Wall Street; 6) demand regular audits of the Federal Reserve, which would increase the transparency of its operations; 7) eliminate restrictions on starting new banks; and 8) permit the creation of new currencies.

Limitations. Arguably, Congressman Paul's approach, though currently limited by a lack of voter support—sustainably minded or otherwise—has a better chance of being actualized than the nationalized approach, due to a political environment still enamored of the free market system. However, Greco views the return to commodity money as a regression to a "more primitive means of exchange" (2009: 99). According to Greco, backing money with a commodity, gold or otherwise, limits the amount of money in the money supply, and thus does not solve the dilemma of how to make money available to all members of the global population. Nor does commodity money discourage hoarding, competitiveness, and wealth concentration. Thus, he believes, "The answer to the abusive issuance and circulation of credit money lies not in turning back the clock and reverting to more primitive forms, but in perfecting the superior form, credit money, within the arena of free competition" (2009: 99).

Additionally, given a centralized monetary and economic system that has so successfully concentrated money and resources in the hands of wealthy individuals, corporations, financial institutions, and governments, a commodity-based monetary system can be manipulated and used to the advantage of those who own, control, or have the means to financially influence the commodity in question. This is particularly true of gold, which is why many monetary reformists oppose the gold standard and prefer a monetary system backed by a basket of commodities, which is far more difficult to manipulate (Lietaer, 2001). Finally, while Paul's approach is focused on creating sound money, it does not include measures for how to provide a means of exchange to those who have been denied participation in the current system. Those who have been left out of the current system require a more equitable and abundant means of exchange in order to foster bottom-up economic development.

The Terra: A Commodity-Based Global Reference Currency

Bernard Lietaer also supports private, free-market approaches for the creation of a sustainable global monetary system. He proposed the creation of a global reference currency (GRC), "which is not tied to any particular nation state, and whose main purpose is to provide a stable and reliable reference currency for international contracts and trade" (2001: 249). Lietaer's proposed GRC is named the Terra, and its unit account is a basket of commodities. The purpose of the Terra is to relink the global monetary system to the natural world, thereby curtailing the speculative global currency casino.

Perhaps the most important feature of the Terra is that it would, unlike Paul's gold-backed monetary system, contain a demurrage fee (sustainability fee). Demurrage charges on currency were devised by the theoretical economist Silvio Gesell and can be thought of as a negative interest rate (Lietaer, 2001). The purpose of the demurrage charge is to increase the velocity, or rate of circulation, of money, thereby systematically discouraging currency hoarding and hierarchical wealth concentration. For the Terra, the demurrage fee is built into the system due to the cost of storing a basket of commodities. In theory, the Terra would temper global economic cycles of boom and bust by operating countercyclical to the business cycle. It would accomplish this task by providing liquidity during times of bust and stability during times of boom. The ultimate goal of the Terra is to redirect the global economy from its primary focus on exponential quantitative growth that is unsustainable, to steady-state qualitative growth that is sustainable.

Limitations. As a private and global venture, the Terra requires the support and participation of the corporate world for its implementation. According to Lietaer, this scenario is feasible because the Terra could provide a benefit to global corporations by mitigating the costs of foreign exchange risks. The Terra, as a geopolitical venture, requires a new Bretton Woods Agreement and/or the reformation of the IMF (Lietaer,

2001). The United Nations, China, and Russia have already suggested that the IMF create a new GRC to replace the increasingly unstable U.S. dollar as the global currency (Bryanski, 2009). This means that the future implementation of a new Bretton Woods Agreement is indeed a possibility. However, in this scenario, the IMF would be called upon to turn its internal unit of account, called Special Drawing Rights (SDR), into a new global currency. SDRs would be backed by a basket of currencies, not a basket of commodities. Thus, the implementation of the Terra as a private venture, though challenging, is a more likely a scenario than the IMF being reorganized to accommodate a commodity-based global currency.

Mutual Credit Systems

In mutual credit systems, participants create credit (money) at the moment of each transaction (Lietaer, 2001). Rather than have their credit privatized and loaned back at interest from a monopolized, politicized, and centralized monetary system, participants create, issue, and govern their own credit in a debt free monetary system. Greco outlined the economic benefits of mutual credit systems as follows: 1) Mutual credit is abundant and balanced, rather than scarce and unstable 2) The allocation of credit is democratically determined by all participants rather than by a central authority 3) Mutual credit is *not* positive interest bearing 4) Mutual credit systems are immune from inflation and, by extension, economic boom and bust cycles and 5) Mutual credit systems are sustainable and engender sustainable and equitable economic development.

Mutual credit systems are founded upon the principle of direct credit clearing. Direct credit clearing allows buyers and sellers to exchange goods and services without the process being governed by a third party's monopolized currency or instrument of credit (Greco, 2009). The goal of mutual credit systems, then, is not to regress to more primitive forms of money, but rather to find a way to reclaim the credit commons from the centralized monetary system, thereby providing an equitable and abundant means of exchange to all participants.

Also a proponent of mutual credit systems, Bernard Lietaer (2000, 2001) believed that an effective mutual credit system must, like the Terra, contain a demurrage fee. A demurrage fee is key because it increases the rate at which currency circulates in a system. So, not only is negative interest bearing mutual credit abundant and self-regulating, its increased velocity allows it to benefit a wider portion of the population, particularly those who have been excluded from participation in the centralized monetary system. Because the demurrage fee discourages hoarding, mutual credit systems have the potential to encourage investment in long-term community investments, thereby fostering bottom-up qualitative economic growth necessary for the creation of an equitable and sustainable world.

Perhaps the most outstanding feature of a negative interest bearing mutual credit system is that it propagates and reinforces sustainable behavior and a sustainable mindset (Lietaer, 2000). Whereas centralized monetary systems promote individualism, competition, hoarding, and the hierarchical accumulation of wealth, resource, and power, mutual credit systems have the potential to promote collectivism, cooperation, community building, sufficiency, and the equitable distribution of wealth and power, all of which are necessary for the creation of a sustainable world. Thus, a negative interest bearing mutual credit system is the only type of monetary system, envisioned thus far, that addresses the issue of how to systematically foster a sustainable mindset.

Presently, only limited, though promising, case studies of the effects of mutual credit systems on cultures of past (Lietaer, 2000), such as Europe during the central Middle Ages and ancient Egypt, and present (DeMeulenaere & Lietaer, 2003), such as Bali and Papua New Guinea, support the claim that mutual credit systems have the potential to engender sustainable behavior and actualize bottom-up, qualitative economic development. Unfortunately, long operating mutual credit systems are few and far between. Nonetheless, a sufficient amount of mutual credit systems exist that can be studied for the purpose of better understanding their psychological and cultural effects. Thus, future applications of the field of psychology to the study of mutual credit systems are crucial for understanding how best to systematically propagate and reinforce sustainable behavior.

Greco outlined three steps for achieving monetary transformation founded upon mutual credit systems. First, create mutual credit clearing exchanges that operate outside the grasp of the existing monetary and financial system. Second, network these exchanges together in a global payment system. And, third, employ a concrete standard of value and unit of account that cannot be "manipulated to the advantage of any particular bank, cartel, government, or individual" (2009: 108). Greco firmly believes that mutual credit systems have the potential to create a peaceful and sustainable worldview. He stated:

> With the proper implementation strategies and the application of sound financial principles, direct mutual credit clearing is an unstoppable technological advance that has the power to ameliorate myriad economic, political, and social problems and establish a solid foundation needed to enable civilization to make a major leap towards a harmonious, sustainable way of living (2009: 108).

Limitations. Mutual credit systems are not a new phenomenon. For example, the WIR in Switzerland has been operating since 1934 (Greco, 2009). Yet, despite the presence of numerous mutual credit systems, such as Time Hours, Ithaca Hours, and the LETSystem, these systems have either failed or operate on a limited basis only. Greco (2009) outlined four factors that limit the success of mutual exchange systems: 1) poorly designed architecture and/or currency, 2) poor or insufficient management, 3) insufficient implementation strategy, 4) inappropriate context or opportunity within which the system was introduced. Additionally, research studies on the motivations and beliefs of mutual exchange system participants revealed that membership is dominated by greens and liberals (Caldwell, 2000). One researcher concluded that the domination of mutual credit systems by greens and liberals limits their ability to attract support from moderates and conservatives (Gran, 1998). Bipartisan participation is crucial for the long-term success of mutual credit systems.

The domination of mutual credit systems by greens and liberals is a problem of implementation. However, that mutual credit systems are superior and evolved monetary systems that provide a fair, equitable, and sustainable means of exchange for all participants, including moderates and conservatives, is not sufficient for the task of attracting mainstream support. Greco stated: "People must be made aware that the superior product exists, they must be persuaded that the advantages of adopting it outweigh the risks and disadvantages, and the product must be made easily available at a price they can afford" (2009: 151). What is needed, according to Malcolm Gladwell, is a social network comprised of "mavens," "connectors," and "salesmen" (as cited in Greco, 2009). Mavens present the message. Connectors take the message out into the world and extend it to as many social groups as possible, not just liberals and greens. And, salesmen persuade other people to buy into the new product, system, or paradigm.

Another critical success factor is the situational context in which the mutual credit system is implemented. Greco pointed out that two of the most successful mutual credit systems, the WIR in Switzerland and Argentine "credito" currencies, were introduced "at a time and place in which conventional money was acutely scarce, banking and financial institutions were in disarray, and the national economy was in distress" (2009: 153).

Though the multi-trillion dollar bailout of Wall Street temporarily staved off economic and monetary collapse, the crisis did increase awareness of the need for monetary reformation. Assuming that the crisis generated sufficient awareness of the unsustainable centralized monetary system as to bring this issue to the forefront of the sustainability movement, now would be a strategic time for the implementation of an entirely new money agreement.

The sustainability movement must come together and design an effective and realistic strategy for the creation of an ethical and sustainable monetary system. This strategy must address the issue of whether to focus on reforming the existing monetary system or on reclaiming the credit commons via the implementation of an alternative monetary system. Finally, this strategy must address the issue of how best to propagate and reinforce the sustainable mindset required to create a peaceful and sustainable worldview. But first, a second wave strategy requires an articulated second wave vision.

A Second Wave Vision: Sustainable Abundance

Of all the sustainable visions put forth since the release of the *Brundtland Report*, Bernard Lietaer's is perhaps the most comprehensive and appealing in that it fosters both sustainable economic development *and* psychospiritual development. Lietaer called his vision "Sustainable Abundance," which he defined as an approach to sustainable development that "provides humanity with the ability to flourish and grow materially, emotionally and spiritually without squandering resources from the future." Lietaer further elaborated:

> It is characteristic of a community, society, country or global system that gives people the opportunity to express their highest creative calling, without diminishing the prospects for coming generations to enjoy the same or a better way of life. It is about having our material needs met so that we can explore our highest potentiality as human beings (2001: 17).

Lietaer was quick to point out that his use of the term "abundance" should not imply the accumulation of superfluous material goods, as is the custom of wealthy individuals and nations within the current worldview. Following is "abundance" as defined by Lietaer:

> Abundance is what provides enough freedom of choice in the material domain to as many people as possible, so that they can express their passion and creativity. Such

creativity is the expression of their highest form of consciousness, their highest calling, and provides a true sense of meaning in their life (2001: 117).

The goal of Sustainable Abundance is to provide the opportunity for humanity to actualize its full potential within an ecologically sustainable global economy. Founded upon a democratically governed and operated complementary currency system based on mutual credit, Sustainable Abundance makes it possible for all people to have a more equitable chance at generating new wealth for themselves, their families, and their communities in a manner that is sustaining of the ecological environment. The polarization of wealth, resources, and power has weakened the vibrancy and vitality of social and ecological support networks to the detriment of all beings. Sustainable Abundance provides a way to mend the schism that separates and alienates people from each other and from the natural environment.

A Second Wave Strategy

The recent economic crisis ignited global concerns about the increasingly unstable U.S. dollar's position as the global reserve currency. In response to this concern, the United Nation's *Trade and Development Report 2009*, called for a new Bretton Woods Agreement that would convert the IMF's SDR into the new global reserve currency. One of the suggested purposes of the new system would be to operate countercyclical to the global economy by providing liquidity during times of economic recession. In addition to a new global reference currency, the UN also called for the creation of an effective system of governance to oversee international monetary and finance regulations. The purpose of the new system of governance would be to help protect the global economy from future economic crises stemming from the instability of a national global reference currency.

Undoubtedly, the UN's concerns are legitimate. However, all of the UN's proposed solutions for stabilizing the global centralized monetary system rely on changes within the existing system, such as implementing a new global reference currency and an international system of governance. Furthermore, the UN's proposal supports the further centralization of the global monetary system by upgrading the role of the IMF to that of a global central bank (Emsden & Parussini, 2009). Nowhere in the report did the UN suggest reclamation of the credit commons via the implementation of mutual credit systems that could help foster bottom-up economic development. If allowed, this further centralization of the global monetary system would make it increasingly difficult to actualize bottom-up economic development that is equitable, sustainable, and inclusive of all members of the global population.

Again, it is beyond the scope of this chapter to suggest a first wave approach for monetary reformation within the centralized monetary system. This approach must come about, after open debate and discussion, via consensus within the sustainability movement as a whole. However, as Greco and Lietaer argued, the most effective approach should most likely come about as a private, free-market initiative. And, though Greco viewed commodity money as more primitive than credit money, it might make sense, in the short term, for the sustainability movement to support the backing of the U.S. dollar and/or a new GRC with a basket of commodities, as this would foster stability by grounding the global monetary system back into the natural world. The sustainability movement would be wise to assume that future economic and monetary crises will result in proposals calling for the increased centralization of the global monetary system via a new GRC issued by the IMF and/or the merger of neighboring national currencies into one currency, as exemplified by the euro. Thus, the strategy of sustainability's second wave must simultaneously strive to both stabilize the existing monetary system in order to counter further centralization *and* create a viable alternative to the dominant monetary system that engenders Sustainable Abundance.

However, change within the existing monetary system, though not impossible, will be difficult to actualize. Thus, in accordance with the wisdom of Audre Lorde and Lewis D. Solomon, the sustainability movement should focus a considerable amount of its time and energy on creating an alternate means of exchange that is inclusive of all people. This chapter suggests that negative interest bearing mutual credit systems are the most comprehensive approach for creating an alternative means of exchange that is ethical, equitable, and sustainable. The justifications for negative interest bearing mutual credit systems are psychological, economic, and strategic.

Negative Interest Bearing Mutual Credit Systems

Psychological Benefits. That the centralized monetary system both demands and reinforces unsustainable individual, collective, and institutional behavior is one of the greatest challenges faced by the sustainability movement. Lietaer recognized the ability of money to dictate matters other than finance and economics with the following passage: "Money is to a civilization as the DNA code is to a species. It replicates structures and behavior patterns that remain active across time and space for generations" (2000: 273). Thus, one of the strengths of negative interest bearing mutual credit systems is that they help propagate and reinforce sustainable behavior.

As was outlined previously, centralized monetary systems promote individualism, competition, hoarding, and the hierarchical accumulation of wealth, resource, and power. In mutual credit systems participants, rather than an imposing hierarchical monetary authority, democratically govern the system and the issue of credit. This results in increased participation and interaction amongst members, as well as between members and their communities. As such, negative interest bearing mutual credit systems promote collectivism, cooperation, community building, sufficiency, and the sharing of wealth and power, all of which are necessary for the creation of a sustainable mindset. Their ability to systematically propagate and reinforce sustainable behavior and a sustainable mindset is foundational for actualizing Sustainable Abundance.

Economic Benefits. Negative interest bearing mutual credit systems are open to all members of society and grant each participant the right to create and govern their own credit. Credit, in these systems, is abundant, balanced, and self-regulating. Due to a negative, rather than positive, interest rate, mutual credit systems do not lend themselves to a debt imperative or a quantitative growth imperative. Instead, they encourage steady-state, qualitative economic growth and long-term investments in such areas as community health and development, maintenance, and infrastructure. In short, negative interest-bearing mutual credit systems make possible the actualization of a sustainable world.

Strategic Benefits. Negative interests bearing mutual credit systems have two important strategic benefits. First, they can be presented as complementary and, thus non-threatening, to the centralized monetary system (Lietaer, 2001). According to Lietaer, negative interest bearing mutual credit systems temper the business cycle by providing liquidity during times of economic crisis. This feature is appealing and might be welcomed, rather than rejected, by a centralized monetary system that is increasingly unstable. Mutual credit systems have a far better chance of being allowed to grow and develop if they are, from the start, diplomatically presented as being a *complement* to the centralized monetary system, rather than an *alternative* to the status quo. This strategy might also help with the process of drawing support for mutual credit systems from moderates and conservatives. And second, as free-market ventures, mutual credit systems do not require the support of a political majority for their implementation. With an orchestrated effort within the sustainability movement, well-designed mutual credit systems can be strategically implemented at any time.

The shift to local, and potentially regional and global, mutual credit systems requires "a shift in our perception of our relationship to money, the first in centuries" (Lietaer, 2001: 20). A large-scale collective shift in consciousness and perception about money is not likely to occur expeditiously during times of economic stability. However, this shift is likely to be accelerated during periods of economic crisis, especially if a viable complementary monetary system is ready and waiting to help guide the way.

As the centralized monetary system is increasingly unstable, it would be wise of the sustainability movement to have a complementary monetary system, comprised of local mutual credit systems, in place before the next, and inevitable, economic crash. The financial elite maneuvers periods of economic crisis for its own advantage, so why should not the sustainability movement utilize future and inevitable economic crises to engender support for a complementary monetary system that is ethical, equitable, and sustainable?

The sustainability movement must be realistic and acknowledge that it might not be able to actualize monetary change within the existing monetary system. Thus, it is paramount that the sustainability movement implement a complementary monetary system before the pressures of peak oil, ecological crisis, natural disaster, and/or global violence and war collide with a global economic bust cycle, thereby creating sufficient instability to bring down the house of cards. Without an alternate system in place, and in the face of tremendous global pressure to further centralize the monetary system, the sustainability movement will have little chance of actualizing Sustainable Abundance.

Conclusion

The course of human development has been dictated by hierarchy and domination to the detriment of social justice and the health of the ecological environment. Hierarchy and domination are perpetuated by a global centralized monetary system that consolidates wealth, resources, and power in the hands of the few at the expense of the many. This arrangement is unethical and unsustainable, and requires the immediate attention of the sustainability movement as a whole. As Greco argued, "the control of money and exchange mechanisms is the key structural element that determines the distribution of power ... and it must be the main focus if any degree of community empowerment and self-determination is to be achieved" (2009: 46).

The challenges we face are overwhelming, but we must remind ourselves that the actualization of Sustainable Abundance is possible. However, the dream of Sustainable Abundance cannot be manifested in reality without the unified support of the sustainability movement. We must also call upon other like-minded movements for their assistance, particularly the peace movement, as the efficacy of both the peace and sustainability movement is hindered by the centralized monetary system. Additionally, we must partner with the field of psychology in order to better understand how sustainable monetary systems contribute to the process by which a sustainable mindset is forged. Sustainable economists have been dabbling in the field of psychology for years. It's time for sustainably-minded psychologists to return the favor. And, finally, the sustainability movement's mavens, connectors, and salesmen must spread the message of sustainable monetary systems to all corners of the world.

Our message should be clear: Money makes a great difference in the lives of people and communities. The justifiable purpose of money is to serve the needs of local, regional, and global communities in a way that honors the limits of the ecological environment. The purpose of money is not to serve the needs of elite power. Therefore, we must not settle for a means of exchange that is still designed to serve the purpose of competitive growth for the most powerful players merely because this elite offers more benevolent, equitable, and sustainable changes than those now in place. Instead, we must join together as a movement and make possible a sustainable future that is empowering of all people, and all communities, in the global village.

References

Bryanski, G. (2009, March 19). China backs talks on dollar as reserve—russian source *Reuters*. Retrieved September 22, 2009 from http://www.reuters.com/article/usDollarRpt/idUSLJ9363302009031 9.

Caldwell, C. (2000). "Why do people join local exchange trading systems?" *International Journal of Community Currency Research*. Vol. 4. Retrieved September 22, 2009 from http://www.uea.ac.uk/env/ijccr/.

Emsden, C. & Parussini, G. (2009, October 5). IMF stakes claim to be everybody's central bank. *The Wall Street Journal*. Retrieved November, 19 2009 from http://online.wsj.com/article/BT-CO-20091005-700491.html.

Gladwell, M. (2002). *The tipping point: How little things can make a big difference*. New York, NY: Back Bay Books/ Little, Brown and Company.

Gran, E. (1998). "Green domination in norwegian letsystems: Catalyst for growth or constraint on development?" *International Journal of Community Currency Research*.

Vol. 2. Retrieved September 22, 2009 from http://www.uae.ac.uk/env/ijccr/.
Greco, T., Jr. (1990). *Money and debt: A solution to the global crisis.* (2nd ed.). Tucson, AZ: Thomas Greco, Jr.
Greco, T., Jr.(1994). *New money for healthy communities.* Tucson, AZ: Thomas Greco, Jr.
Greco, T., Jr. (2001). *Money: Understanding and creating alternatives to legal tender.* White River Junction, VT: Chelsea Green Publishing.
Greco, T., Jr. (2009). *The end of money and the future of civilization.* White River Junction, VT: Chelsea Green Publishing.
Hawken, P. (2007). *Blessed unrest: How the largest social movement in history is restoring grace, justice, and beauty to the world.* New York, NY: Penguin Group.
Klein, N. (2007). *The shock doctrine: The rise of disaster capitalism.* New York, NY: Metropolitan Books.
Korten, D. (2009). *Agenda for a new economy: From phantom wealth to real wealth.* San Francisco, CA: Berrett-Koehler Publishers, Inc.
Lietaer, B. (2000). *The mystery of money: beyond greed and scarcity.* Munich: Riemann Verlag.
Lietaer, B. (2001). *The future of money: Creating new wealth, work, and a wiser world.* London: Century.
Lietaer, B. & DeMeulenaere, S. (2003). Sustaining cultural vitality in a globalizing world: The Balinese example. *International Journal of Social Economics, 30*(9). Retrieved November 22, 2009 from www.appropriate-economics.org.
Paul, R. (2009). *End the fed.* New York, NY: Grand Central Publishing.
Pilisuk, M., & Gaddy, E. (in press). Against so much money and power, can the peace movement succeed?. In M. N. Nagler and M. Pilisuk (Eds.), *Peace movements worldwide: History, psychology, and practices*: Vol. 1. *History and vitality of peace movements*. Westport, CT: Greenwood/Praeger Publishers.
Pyszczynski, T., Solomon, S., & Greenberg, J. (2003). *In the wake of 9/11: The psychology of terror*. Washington, D.C.: American Psychological Association.
Solomon, L. (1996). *Rethinking our centralized monetary system: The case for a system of local currencies*. Westport, CT: Praeger Publishers.
The World Commission for the Environment and Development (1987). *Our common future*. Oxford, England: Oxford University Press.
United Nations Conference on Trade and Development. (2009). *Trade and development report*. Retrieved November 19, 2009 from http://www.unctad.org/.
Van Gelder, S. (ed.). (2009). The new economy starts here: Why this crisis may be our best chance. *Yes!*, 50.
Zarlenga, S. (2002). *The lost science of money: The mythology of money—the story of power*. Valatie, NY: American Monetary Institute.

Economic Genesis: The Evolution of a Sustainable Economy through Values-Based Social Architecture

Marjorie Kelly

Introduction

"On or about December 1910, human character changed," Virginia Woolf once observed. She was writing in the early 1920s, as a wave of Modernist transformation was sweeping the globe in the wake of World War I and the dashed gentility of the old world it left on the battlefield of Verdun.

One might make a similar observation about our own time, as we see the subprime mortgage crisis becoming our own Verdun, leaving in its wake a new form of dashed gentility. The proud tower of Wall Street was seen sagging on its foundation on September 16, 2008, as the *Wall Street Journal* reported on page one, "Wall Street as we have known it is ceasing to exist." In the days surrounding this pronouncement, many of the nation's major investment banking firms were crashing in puffs of dust (one entering bankruptcy, two bought for a pittance, two transforming themselves into regulated banks), like so many dominos falling in the slow-motion explosion of the credit crisis. The collapse of two of these venerable old firms followed by only a few days the nationalization of mortgage giants Fannie Mae and Freddie Mac, as the U.S. government took on their $5.4 trillion in loans, roughly equal to the entire publicly traded federal debt. With barely a pause for breath, regulators engineered a bailout for the largest insurer in the world, American International Group, and orchestrated a $700 billion bailout to be spent sopping up bad investments worldwide. Amid the chaos, the growing geriatric fragility of what was once the largest corporation in the world, General Motors—at the time hemorrhaging billions—proved barely a footnote.

Yet even when taken as a whole, these outsized Wall Street events amounted to little more than a punctuation mark in a larger sentence about massive change that had begun less than a year earlier with Al Gore's Nobel Prize for his global warming work, unfolded into a long clause about $150-a-barrel oil, and ended with the parenthetical remark of Hurricane Ike hurtling itself upon Galveston, Texas on the unlucky day of September 13. In the background, there were reports about the decade-long drought in Australia, the water shortages in Africa and Asia, the unexpectedly rapid disintegration of Antarctica's vast Wilkins ice shelf, food riots in a dozen nations, and the ongoing die-off of salmon and amphibians and much-beloved birds.

We stand today at a hinge moment. A massive shift is underway in our economy and on our planet. Looking back on our time a decade or two hence, it may not prove hyperbole to observe: On or about September 2008, the character of the modern economy changed.

Signals of Large-Scale Transformation

It is in the language of crisis that transitions announce themselves. As old patterns dissolve and new ones form, turbulence arises. What we are experiencing today is a systemic crisis, heralding the beginning of a large-scale transformation of the deepest structures of cultural organization. In systems thinking, transformation is distinct from adaptation, which is the normal process of incremental adjustments in policies and institutions. Transformations are something else. They are those rare moments in history when dominant institutions become unfit for an unfolding new set of circumstances and change in profound ways.[1]

We're experiencing today a system event operating at the scale of the planet. We are on the threshold of a passage likely to prove as significant as the advent of agriculture and the dawn of the Industrial Revolution. Such a transformation inevitably wears many faces—geo-political, cultural, technological—but at the most fundamental level, it involves a transformation in the ecological/economic order. In our industrial-age mindset, we think of ecology and economy as separate realms. But their Greek origins indicate their kinship. Ecology is from *oikos* (house) and *logos* (logic or story), while economy is from *oikos* and *nomos* (law or convention). Together they point to a single phenomenon: the organizational design and operating principles of a household.

The household is the Earth. We as human beings have grown, willy-nilly, into the role of its householders and stewards. That we are massively inept in this role is the message being telegraphed to us from the unfolding ecological and financial chaos, which is emanating from our own flawed operating principles, which are in turn imbedded in flawed institutional designs.

The economic designs we rely upon are out of synch with 21st-century reality. We see these outmoded patterns of organization, for example, in investment banks that see credit-default swaps as a safe foundation on which to rest the global economy. We see these patterns in corporate balance sheets that treat rivers and the air as having zero value, while every crate of nuts and bolts is numbered with precise zeal. We see these designs in federal policies fixated on GDP, which treats a dollar spent on toxic cleanup the same as a dollar spent on educating a child.

It is clear that this old system design is inadequate to the times. Our immediate tasks are also clear: to redesign the regulatory framework of finance, and to rebuild our energy infrastructure for sustainability. These are not small tasks. But neither are they patchworks upon a frame that is essentially sound. The larger challenge we face involves the purpose and design of the capitalist economy.

The Evolution of Social Architecture

Do economies exist to create wealth for the financial elite, or to serve broad human well-being? This is the question the current crisis brings to the fore. It is a question of values and norms, and as such it is intertwined with institutional design. Institutions are the most powerful and pervasive elements of modern society and the most important carriers of cultural norms.[2]

Richard Nelson, an economics professor at Columbia University who co-founded the field of evolutionary economics, observes that social systems evolve because of two kinds of innovation: advances in physical technologies (such as new environmental and energy technologies), and advances in social technologies (such as new forms of organization). Much attention today is on physical technologies needed for sustainability. But social technologies are equally as fundamental. The two must co-evolve.[3]

Another term for social technologies is *social architectures*. The most basic of these in our economy are the corporation and the capital markets. They are central to the framework of the economy as a living system, shaping its energy flows. Deep frameworks like these are highly resilient.

Up to a point. When they encounter disturbances beyond a critical threshold, they can transition to a qualitatively different state. Tolerable stresses lead to adaptation. Destabilizing stresses lead to transformation.[4]

In living systems, transformation occurs through self-creativity. "Natural systems evolve new structures and new functions; they create themselves in time," writes Ervin Laszlo in *The Systems View of the World.* Beyond the ongoing process of *ontogenesis*—the growth of the young—there is the broader process of *phylogenesis,* the evolution of species. This is "the creative advance of nature into novelty: it is the trailblazing self-transformation of entire species and populations of organisms," Laszlo says. In natural systems, behavior is a result of structure. Thus, transformation involves an evolution in a system's core structures.[5]

As destabilizing stresses mount today, new species of economic designs are at the same time rapidly emerging. We see this in carbon trading schemes, new kinds of stock exchanges with social listing requirements, nonprofit-owned "social enterprises" that operate profitably while pursuing social mission, the burgeoning microfinance industry, community development financial corporations, community-supported agriculture, and emerging hybrids like "for-profit philanthropy."

In experiments large and small around the world, novel structures are dissolving the old boundaries of organizational forms. These new designs represent potential models of social architecture for a sustainable future—models that could prove better adapted to a new era. Grameen Bank, which pioneered the practice of making tiny loans that are made to aid the poor, was largely shielded from the credit crisis, because its borrowers were also its depositors—insulating it from the larger turbulence. Similarly in the U.S., credit unions—like Self-Help Credit Union, a leader in combating predatory lending—have remained more financially sound than large, traditional banks.[6]

Organizations like these are resilient because they are self-organized around serving the needs of life, not around speculative booms that go bust overnight. They offer hints of the design principles we will need in the coming years, as we live out the answers to two questions: *How can we make our economy a living one?* And because economic institutions shape an economy, answering this question will require us to answer a second: *How can we design economic institutions that are life-sustaining?*

Our first inclination is toward mechanisms of hierarchy and control: constraining the abuses of the financial system, controlling carbon emissions, and so on. Yet in a broader evolutionary frame, emergence is how deep change happens. Small-scale experiments arise. These become connected through networks, which strengthen into communities of practice. Supported by government policies and the evolution of cultural norms, a new system can ultimately emerge at a greater level of scale.[7]

Scenarios of the Future

We are today in a liminal time, in the doorway of great change but not yet fully entered upon a new social order. As old social architectures break down, the future remains disturbingly unformed. To explore what might lie ahead, in the 1990s my colleagues at the Tellus Institute developed a half-dozen future scenarios. Today many think our options are narrowing, with one colleague suggesting we face a knife-edge phenomenon. We are poised on the narrow edge of the present and could ultimately tip into one of two likely futures.

One is a kind of Fortress World, a landscape of devastation where the affluent come to live in protected enclaves amid oceans of misery. The other is a world of a Great Transition. Here, we do not succeed in staving off all effects of climate change—droughts, hurricanes, millions of climate change refugees—but these serve to deepen our understanding of interdependence. Our culture begins, in its soul, to grasp that all life on earth is deeply intertwined. With this new sensibility, an unprecedented economic system emerges that is both more humane and better adapted to a new world of limits.[8]

No one can say what the future will hold, not because we lack the foresight to see ahead, but because there's nothing out there to see. The future isn't lying ahead on a line beyond past and present, unscrolling inevitably forward. It literally does not exist. Until we create it, out of the mix of accidents and forces and crises and the way we respond to them. And the dreams we bring to them. And the hope we allow ourselves to harbor, against all practical reason.

The critical difference between a Fortress World and a Great Transition is one of cultural spirit. A spirit of fear. Or a spirit of hope. It is our spirit that will allow us to transform the inevitable crises ahead in positive ways. Which scenario unfolds will depend upon the cultural ethos we both create and inhabit: an ethos that says each of us is out for ourselves, or we're all in this together.[9]

The Central Role of Values

Social systems are value-centric. They self-organize around what people care about: the things we value. The values of *liberty, equality,* and *justice* represent the genetic strands of our political order, the central organizing principles of our governing institutions. Our economy is organized around the values of *individualism, growth,* and *free markets.* Growth in wealth and material goods is perceived as the way to feed individual prosperity. Free markets create the running room in which individuals can pursue their interests.

For many today, these values are beginning to feel incomplete or illegitimate. Americans left, right, and center share a moral anxiety that selfishness is spiraling out of control.[10] There is a sense that we "live in a world tranquilized by trivia and adrift in drivel," wrote Rabbi Michael Lerner in *The Left Hand of God.* As Lerner found in a series of interviews, many middle Americans say their living energies are being depleted, that life is about going through motions imposed by forces beyond their control. Many hunger for work and lives of greater purpose and meaning.[11]

The unacknowledged spiritual crisis we face has been wrought by the growth of an amoral economy. It is a sign the economy's moral foundation is eroding. When institutional behavior veers far from people's sense of right, it calls institutional legitimacy into question. Previous ethical crises in business have been dismissed as the work of a few rogue executives. Today we have no such luxury. When the inflated price of houses in Phoenix and Miami sends shock waves through banks in Germany and Iceland, the crisis is systemic. It is not about the unethical behavior of errant individuals but about the legal and indeed mandatory profit-seeking behavior endemic to the system itself.

Ethical codes and laws will always remain critical, but many of those we have created thus far function essentially as picket fences, seeking to surround but not change the value system at the heart of economic institutions. What the financial crisis offers is an opportunity to see how the system's central values lead directly to trouble.

Individualism shorn of responsibility to the whole is the amoral motive at the root of the crisis. As New Deal banking regulations became antiquated, capital markets innovated around them. What arose was a shadow banking system in hedge funds, private equity firms, and investment banks. Because regulations failed to touch the system's primary driving force, the energies in the system simply flowed around the regulations—like water around a boulder.

The core aim was making as much money as fast as possible, and externalizing the consequences onto others. It was there in mortgage brokers intent on collecting commissions, indifferent to the fact that clients were being put into abusive mortgages they couldn't afford. It was there in investment banks that collected enormous fees for repackaging these mortgages into collateralized debt obligations, not asking how sound the lending was. It was there in the institutions that bought these instruments, seeking to protect themselves with credit default swaps, which meant someone else would pay if the debts went bad.

To the capitalist mind, it was utopia: an infinite universe of new financial products to trade, limitless potential gains, and zero downside risk. But what was left out was systemic risk. It is impossible for every individual to offload risk on someone else, because then the risk becomes diffuse, spread throughout the system. It is like each of us throwing our trash away in a neighbor's yard. In an interdependent social system—just as in an ecosystem—there is no "away."

What we are witnessing, writ large, is a demonstration of the fact that individualism, growth, and free markets are flawed values around which to organize a global economy. Morality and reality are met in this truth.

An Emerging Suite of Sustainable Values

As the illegitimacy of current economic values stands exposed, a new suite of values is quietly emerging: *interdependence, sustainability,* and *well-being.* We see these in the growth of organic foods, demand for Prius autos, the popularity of books like *In Praise of Slow.* These emerging values do not so much replace the old value system as complement it, bringing to it a greater complexity.

Interdependence is at work in the credit crisis, when the default of poor families deceived into mortgages they can't afford threatens the pyramid of collateralized debt obligations which in turn leads the global credit system to stall. Interdependence is similarly at work ecologically, when U.S. automobile use creates demand for ethanol that drives up the price of grain in Haiti and sends a child to bed hungry. The economic web is the social counterpart of the ecosystem. It is said that the flapping of butterfly wings in Sao Paulo can lead to a hurricane in Miami. The same is true of a sales commission unfairly earned, or a contract signed under coercion, which can lead to a hurricane in global financial markets.

Interdependence is the nature of life. This is the truth religions have taught for millennia. The Jewish tradition speaks of *tikkun olam,* the duty each of us has to heal and repair the world. The Christian tradition admonishes love of one's neighbor as oneself. In Islam, the Koran says the tribes have been created to know one another. In the Buddhist worldview, the separate, isolated self is an illusion. This is also the systems view, which sees all living systems as sub-systems of the Earth. We are one with each other and with the planet that gives us life. We are constituted by, and help to constitute, the living order around us.

Interdependence does not mean losing the integrity of the self in being of service to the whole. Rather, each of us becomes fully who we are only within the context of relationships, communities, and a healthy social order. At the same time, the whole is valued not simply as an end in itself, but because it creates the conditions that preserve the well-being of each person.[12]

Closely related to interdependence is the drive for sustainability. The goal is not the private benefit of isolated individuals, narrowly defined in monetary terms, but the broader well-being of ourselves and the family of life—a well-being embodied in the good life in all its diverse forms: time with friends and family, time for rest, slow food.

The pursuit of well-being, many studies show, generally leads to greater satisfaction than the relentless pursuit of wealth. Psychologist Tim Kasser offers evidence, for example, that people who focus on materialistic values have lower psychological health than others. He emphasizes that our genuine needs are for security, efficacy, connectedness, autonomy and authenticity. Yet the single-minded pursuit of wealth and status leads away from these satisfactions. When people invest heavily in materialistic pursuits, intimacy and connection elude them. Instead of feeling empathy with others, they feel competitive. Rather than feeling free and autonomous, they feel pressured and anxious. The end result is lower vitality and life satisfaction—and in some cases, greater drug and alcohol abuse, narcissism, and antisocial behavior.[13]

Transitioning to a new economy does not mean having less. It means having more of what really matters, which is happiness.[14]

The Mania of Accelerating Growth

If a new economy built on well-being, sustainability, and interdependence seems impossibly idealistic, we might remind ourselves that the truly impossible vision is of infinitely accelerating growth. It's difficult to grasp how much economic growth has kicked into overdrive in recent decades. But consider: Of all the economic output that has occurred worldwide since the time of Caesar, 70 percent has occurred since 1950. If growth in the next 50 years continues at a 2 percent pace—half the annual average since 1950—global economic activity would be *four* times what it is today.[15]

That activity will occur on our one planet. Where half the world's forests and wetlands are already gone. Where three out of four marine fisheries are decimated. If the Earth is to survive four times as much economic activity, what kind of economic institutions should be conducting that activity? Should they be guided primarily by profits?

Profit is unmistakably the driving force behind growth. Since the 1980s, that growth has been kicked into high gear by a process called "financialization." We hear often about globalization, but financialization is of equal import. It means that moving money around now claims a larger share of GDP than making things. In terms of profits, the shift is more dramatic. In 2004, manufacturing brought in less than 10 percent of all corporate profits, while the financial sector inhaled a massive 40 percent.[16]

This process of financialization created an enormous financial bubble that came to overshadow the real economy. Financial assets—such as mortgages, loans, stocks, and bonds—represent claims against real things. Not long ago, these claims, or assets, were roughly equal to global GDP: one dollar of assets for one dollar of income flows. But before the financial melt-down in late 2008, financial assets soared to nearly *four times* global GDP. And financial derivatives—which represent unregulated bets against financial assets—had notional values of more than *ten times* global GDP. It was a house of cards built on top of an asset bubble.

"Think of an inverted pyramid," Charles Morris wrote in *The Trillion Dollar Meltdown*. "The more claims are piled on top of real output, the more wobbly the pyramid becomes. And when large, wobbly objects tumble, they go very fast."[17]

When the air goes out of a financial bubble many times the size of the real economy, it is not like the bursting of, say, the tech stock bubble. The credit bubble encompassed not one industry but the entire global economy. "Credit is the air that financial markets breathe," Morris wrote, "and when the air is poisoned, there's no place to hide."

Just as there is a finite ecosystem that can support only so much economic activity, there is a tangible human economy that can support only so many financial claims. In the mania of financialization, we exceeded the financial carrying capacity of the economy. We are now reaping the consequences.

A New Human Story

Crafting an immediate response to a hurricane takes certain kinds of solutions: food, medicine, rescue efforts. Changing the conditions that give rise to hurricanes requires a larger transformation of the energy system. Similarly, crafting a response to the financial hurricane starts with bailouts and the re-regulation of finance. But ultimate solutions mean a deeper transformation.

We are at a time of coming back to Earth, after a long flight of fantasy about human omnipotence. We are in a time of economic genesis—a time of creating new species of social system design, and a time of writing a new human story of who we are, where we come from, and where we are going. We are in need of a new story about the nature of the good life and the good society, and about the nature of human economic activity.

We have lived long with a worldview that says each of us is out for ourselves. It's not possible to have enough. It's a dog-eat-dog world. We must leave our values at the door when we report for work or search for investments. This is the worldview of fear made incarnate. It is time to ask what an economy based on a worldview of hope might look like.

What cultural shape would the values of interdependence, sustainability, and well-being take, if they were made incarnate in economic institutions? They might take the shape of the cooperative ownership structure of the electricity industry in Denmark, which has helped that nation transition to wind power more quickly than other nations. Or the shape might be like the best of the microfinance industry, where institutions see their aim as helping the poor rather than making a killing off them. It might look like emerging experiments that grant people ownership rights in marine fisheries through "catch shares," which have been shown to halt or reverse catastrophic declines in fish stocks.[18] It might look like a new stock exchange in Brazil, which makes social issues a part of listing requirements.

Institutions like these represent an alternative view of what economic activity can be. And such institutions are not only small and fledgling. Northern Europe has long had a cadre of foundation-owned corporations, such as Novo Nordisk. This pharmaceutical company with 5.6 billion euros in revenues has a mission of defeating diabetes. Its structure requires company leaders to report annually to the foundation board on progress in ensuring that operations are environmentally sound and socially fair. And its stock is publicly traded.

There are also large and successful employee-owned firms, like the John Lewis Partnership in London. It is the largest department store chain in the UK—with revenues larger than those of Monsanto—and is 100 percent owned by its 60,000 employees. It sees its purpose as serving the happiness of its employee partners.

Similarly in the U.S., we find cooperatives like Organic Valley in Wisconsin, a $433 million company owned by the 1,200 organic family farms that produce its milk, cheese, and eggs. Farmer-owners set their own pay rates. And when the company grows, it means additional acres being cleared of chemical fertilizers, additional watersheds washing clean of Roundup and Lasso, and the advent of financial health for formerly stricken farm families.

A New Organizational Archetype: For-Benefit Design

The emerging body of enterprises described here represents a new organizational archetype. They are not entirely profit-driven like for-profit firms. Nor are they exclusively mission-driven like non-profits. They are a hybrid of the two. They've been dubbed "For-Benefit" enterprises, for their purpose is to benefit life, and they aim to be financially self-sustaining—profitable—as they do so. This dual mission lies at the center of their structural design. It's in their DNA. It is imbedded in the defining structures that give each enterprise its shape: why it exists, who owns and controls it, how it is governed, how it is capitalized, and how people are compensated.[19]

The scattered experiments mentioned here may not be ready, any time soon, to replace the major corporations of today. But there is much that can be done to support their growth. In addition to restraining the worst of existing practices, public policy can encourage the growth of healthier alternatives. It can fund community development financial institutions that do fair lending. It can support employee owners in taking over firms when owners retire. It can use government equity stakes in troubled banks as an opportunity to restructure them along for-benefit lines.

In the longer run, how a new system will emerge is impossible to predict. What we can know is that For-Benefit designs have a broader significance than their size would warrant. What they represent is the potential evolution of the capitalist system.

Beneficial economic design is not essentially a technical or legal exercise. It begins with our conceptual picture of *what a company is.* In the dominant paradigm, companies are the *property* of shareholders, devoted to serving their interests. In the emerging paradigm, companies are *living systems.*

This shift works a Copernican revolution, displacing corporations as the center of the universe, making them a sub-system of the larger human and natural order. The needs of this larger living system come first.

When we think of economic institutions as living systems, it allows us to see that this is what they are. They are communities. Central to their design is defining who is a member of the community. There are communities of investors, like traditional firms. Communities of employees, like employee-owned firms. Communities of customers, like credit unions. And multi-stakeholder communities, like some emerging hybrids.

In viewing all economic institutions as the property and playground of wealth, capitalism is shot through with *capital bias,* the notion that persons with wealth have more rights. This is an economic variant of *anthropocentrism*, the notion that human beings have more rights than other species. Both embody the flaw of the old story of Genesis: the idea that human beings have dominion over the Earth and its creatures—and some human beings have dominion over others.

Society has encountered this worldview before, in the social architecture of the monarchy. This ancient and global institution came undone in a mere decade. The process was just getting underway "on or about December 1910," when Virginia Woolf penned her famous observation. The Ottoman Empire had been the first domino to fall, in 1908, followed within three years by the toppling of the Ch'ing Dynasty in China. By the end of the Great War in 1918, three European dynasties lay in the dust: the hapless Romanovs in Russia, the Habsburgs in Austria-Hungary, and the Hohenzollerns in Germany. The British throne would stand, but its imperial possessions would break away, its aristocracy decline into disintegration and disarray.[20]

In our era it is investment banks that have been toppling like dominos, and the dust has yet to clear. There is possibly only one certainty we can have: at such a hinge moment, the actions we take will shape the future for generations to come.

Notes

1. See Paul Raskin, *The Great Transition: The Promise and Lure of the Times Ahead,* Boston: Tellus Institute, SEI Polestar Report No. 10, 2002, www.tellus.org/Documents/Great_Transition.pdf.
2. Heather A. Haveman (Cornell University) and Hayagreeva Rao (Emory University), "Structuring a Theory of Moral Sentiments: Institutional and Organizational Coevolution in the Early Thrift Industry," *The American Journal of Sociology,* Vol. 102, No. 6, May 1997, pp. 1606-1651.
3. R.R. Nelson, 2003, "Physical and Social Technologies and Their Evolution," Columbia University working paper, available from the author; cited in Eric D. Beinhocker, *The Origin of Wealth,* Boston, Mass.: Harvard Business School Press, 2007, p.15.
4. For a discussion of a systems approach to large-scale cultural change see Paul Raskin, "World Lines: A Framework for Exploring Global Pathways," *Ecological Economics, Vol.* 65, April 15, 2008. http://www.tellus.org/documents/PDF/ECOLEC3069.pdf.
5. Ervin Laszlo, *The Systems View of the World: A Holistic Vision for Our Time,* Cresskill, NJ: Hampton Press, 1996, p. 40.
6. Jim Wise, "Prudent Lender Prevails Amid Crisis," *Durham (N.C.) News,* Aug. 16, 2008.

7. Margaret Wheatley and Deborah Frieze, "Using Emergence to Take Social Innovation to Scale," The Berkana Institute, 2008, www.berkana.org.
8. See *The Great Transition: The Promise and Lure of the Times Ahead,* www.tellus.org/Documents/Great_Transition.pdf. Also available are 16 papers on the dimensions of a Great Transition shift; www.GTInitiative.org.
9. For a culture of fear vs. a culture of hope, see Michael Lerner, *The Left Hand of God: Healing America's Political and Spiritual Crisis,* San Francisco: HarperCollins, 2006. The notion of "you're on your own" (YOYO) vs. "we're in this together" (WITT) was articulated by Jared Bernstein in *All Together Now: Common Sense for a Fair Economy,* San Francisco: Berrett-Koehler, 2006.
10. David Callahan, *The Moral Center,* New York: Harcourt, Inc., 2006, pp. 5, 9.
11. Michael Lerner, *The Left Hand of God,* San Francisco: HarperCollins, 2006.
12. Karyn L. Lai, "Conceptual Foundations for Environmental Ethics: A Daoist Perspective," *Environmental Ethics,* Fall 2003, pp. 247-266.
13. Tim Kasser, *The High Price of Materialism,* Cambridge: MIT Press, 2002.
14. John Stutz and Erica Mintzer, "The Affluence Paradox: More Money Is Not Making Us Happier—A Review of Statistical Evidence," Boston: Tellus Institute, June 2006; www.tellus.org/activities/wellbeing.html.
15. This analysis of historical growth draws on the work of Tellus colleague John Stutz and his unpublished working paper, "The Tellus Scenarios in Historical Perspective," 8-21-07 draft, available from the author, JStutz@Tellus.org. Output is real (inflation-adjusted) Gross Domestic Product, measured in purchasing-power-parity-adjusted 1990 dollars. Raw data is from Angus Maddison, *The World Economy: A Millennial Perspective,* Paris, France: Organization for Economic Co-operation and Development (OECD), 2001; and Maddison, *The World Economy: Historical Statistics,* Paris, France: Development Centre of the OECD, 2003.
16. Kevin Phillips, *American Theocracy: The Peril and Politics of Radical Religion, Oil, and Borrowed Money I the 21st Century,* New York: Viking, 2006, pp. 265-267.
17. Charles R. Morris, *The Trillion Dollar Meltdown Easy Money, High Rollers, and the Great Credit Crash,* New York: Public Affairs, 2008, pp. xii-xiii.
18. Cornelia Dean, "Privately Owned Fisheries May Help Shore Up Stocks," *New York Times,* Sept. 19, 2008.
19. See "The Emerging Fourth Sector," working paper by Heerad Sabeti, co-founder of the Fourth Sector Network, www.fourthsector.net. Sabeti coined the term "For-Benefit."
20. Marjorie Kelly, *The Divine Right of Capital,* San Francisco: Berrett-Koehler Publishers, 2001.

References

Bernstein, J. (2006). *All Together Now: Common Sense for a Fair Economy.* San Francisco: Berrett-Koehler.

Callahan, D. (2006). *The Moral Center.* New York: Harcourt.

Dean, C. (2008). "Privately Owned Fisheries May Help Shore Up Stocks," *New York Times,* Sept. 19.

Haveman, H., and Rao, H. (1997). "Structuring a theory of moral sentiments: Institutional and organizational coevolution in the early thrift industry," *The American Journal of Sociology. Vol. 102,* 6, pp. 1606-1651.

Kasser, T. (2002). *The High Price of Materialism,* Cambridge: MIT Press.

Kelly, M. (2001). *The Divine Right of Capital.* San Francisco: Berrett-Koehler.

Laszlo, E. (1996). *The Systems View of the World: A Holistic Vision for our Time.* Cresskill, NJ: Hampton Press.

Lai, K. (2003). "Conceptual Foundations for Environmental Ethics: A Daoist Perspective," *Environmental Ethics,* Autumn Edition.

Lerner, M. (2006). *The Left Hand of God: Healing America's Political and Spiritual Crisis.* San Francisco: HarperCollins.

Maddison, A. (2001). *The World Economy: A Millennial Perspective.* Paris, France: Organization for Economic Co-operation and Development (OECD).

Maddison, A. (2003). *The World Economy: Historical Statistics.* Paris, France: Development Centre of the OECD.

Morris, C. (2008). *The Trillion Dollar Meltdown: Easy Money, High Rollers, and the Great Credit Crash.* New York: Public Affairs.

Nelson, R. (2003). "Physical and Social Technologies and Their Evolution," Columbia University working paper, available from the author; cited in Eric D. Beinhocker's, *The Origin of Wealth,* Boston, Mass.: Harvard Business School Press, 2007, p.15.

Phillips, K. (2006). *American Theocracy: The Peril and Politics of Radical Religion, Oil, and Borrowed Money in the 21st Century.* New York: Viking.

Raskin, P. (2002). *Great Transition: The Promise and Lure of the Times Ahead.* Boston: Tellus Institute SEI Polestar Report no. 10, available at www.tellus.org/Documents/Great_Transition.pdf.

Raskin, P. (2008). "World Lines: A Framework for Exploring Global Pathways," *Ecological Economics, Vol. 65.* http://www.tellus.org/documents/PDF/ECOLEC3069.pdf.

Sabeti, H. (2008). "The Emerging Fourth Sector," working paper, the Fourth Sector Network, www.fourthsector.net.

Stutz, J. Unpublished working paper, "The Tellus Scenarios in Historical Perspective," 8-21-07 draft, available from the author, JStutz@Tellus.org.

Stutz, J., and Mintzer, E. (2006). "The Affluence Paradox: More Money Is Not Making Us Happier—A Review of Statistical Evidence," Boston: Tellus Institute, June. www.tellus.org/activities/wellbeing.html.

Wheatley, M., and Frieze, D. (2008). "Using Emergence to Take Social Innovation to Scale," The Berkana Institute. www.berkana.org.

Wise, J. (2008). "Prudent Lender Prevails Amid Crisis," *Durham (N.C.) News,* Aug. 16, 2008.

From Social Engineering to Community Transformation: Amul, Grameen Bank, and Mondragon as Exemplar Organizations[1]

Dharm P.S. Bhawuk, Susan Mrazek, and Vijayan P. Munusamy

Introduction

Engineering is an applied field that attempts to solve problems by using the theories and methods of science, and indeed it is dedicated to finding technological solutions to human needs, and in that sense is a discipline focused on problem solving. For example, using the knowledge of physics and chemistry, engineers are able to construct tall buildings, providing housing for many people in a limited space; applying the knowledge developed in genetics, they modify food crops that are resistant to certain viruses or yield higher quantity of produce for the same input; employing theories of fluid flow, they develop water systems for irrigation or stents that allow free flow of blood in the hearts of people whose arteries are clogged. It takes years of training to become an engineer, and with increased specialization, even a master's degree is often not sufficient to have an appreciation of the theory and methods of a branch of engineering, and learning a discipline is increasingly becoming a life-long commitment. Though it may sound extreme and belittling, it may not be an exaggeration to say that engineering is a problem solving profession, a noble profession, yet limiting in many ways.

There are some limitations to the engineering approach. First, it necessarily exists in a problem space, and if there is no problem, engineering as a profession would become defunct. Though practically it appears that there will always be problems, it is a theoretical possibility to be in a state where there are no problems. This would become clear when we pause and reflect on the concept of problem, and accept that by nature it is socially constructed. In fact, what is viewed as a problem in one culture is not so in another, and that is why engineers are working on different problems in various parts of the world. Second, it is driven by experts who have years of training, rather than those actually struggling and living with the problem. One of the major limitations of such an expert driven approach is that problems may be defined in a distorted way to fit the knowledge base of science or the expertise of the problem solver. The above limitations have serious consequences when we approach social problems with the zeal of engineers.

When we borrow the engineering model and employ it in social sciences, per force we become social engineers,[2] and borrowing the theories and methods from engineering, we box ourselves in identifying and solving social problems. The approach is necessarily top-down, where experts dictate what a problem is and how it should be solved (Esman & Bruhns, 1966; Esman, 1972). And it is not unusual to find solutions that are of no value, because either they did not solve the problem as perceived by the end users, or they created other problems, even worse than the original malady. We find many examples of how Western experts have exported their solutions, technological and social, to developing countries, with a complete disregard to indigenous solutions and issues (Arisaka, 2003; Bhawuk, 2008) that lead these societies to be worse off than they were before these interventions were implemented.

For example, when a new technology is imposed instead of integrated in the community with the participation of the people, it often destroys the community. The plight of the Yir Yoront aborigine tribe in Australia is a well-known example from anthropological literature, and it all started with the introduction of steel axes by the missionaries toward the end of the nineteenth century. The missionaries were well meaning experts who worked from their own worldview and were trying to help the natives, but instead it "led to confusion, resentment, and general unhappiness in the daily lives of the people; it turned out to be no liberation at all for the Yir Yoront. The failure resulted from an insufficient understanding of the value-laden nature of a particular technology and its total embeddedness in the cultural system of the Yir Yoront" (Arisaka, 2003: 245). Participation from the community could have avoided such a tragedy, and what is instructive is that even the kindness of well meaning missionaries, when coming from the top-down expert-driven mindset, is simply destructive. A similar situation arises when native wisdom is totally disregarded. For example, fascinated by the idea of running water, and oblivious to the constraints of the ecology, the water consumption habits of people in many developing countries are magnified to unnatural levels, and indigenous water systems are neglected to ruins, leading to a chaotic situation of extreme shortage of drinking water. Nepal, a country rich in hydro-power, stands at the brink of water shortage of catastrophic proportion today, and the native wisdom of water management is buried so deep under the engineering debris that people are not sure about ever meeting their water needs.

There is a way out, and it is not to condemn engineering, science, the scientific method, or institution building, but to build on the human strengths and the ability of human beings as individuals and communities to transform themselves. Whereas engineering is about problem solving, human experience is about community building and transformation. In this chapter we discuss how lessons learned in community psychology suggest that we move away from the engineering problem-solving paradigm and adopt the community transformation paradigm, which will allow us to define our needs vis-à-vis our ecology and availability of resources, and to grow together as a community instead of leading an individualistic life that constantly converts communities into markets that cannot meet mindless human desires. We begin by presenting a model derived from the community psychology literature that suggests that there are five essentials of community transformation—the role of the social change agent in the community, the way the problem is defined, information collected, solution implemented, and the need for flexibility in each of these steps. We then present three case studies—Amul (India), Grameen Bank (Bangladesh), and Mondragon (Spain)—that support the model. We conclude the chapter with a discussion of the need to move away from the engineering problem-solving paradigm for social change, and discuss some implications for leadership, sustainable institution building, and organizational development.

Essentials of Community Transformation

Community psychology has come a long way in dealing with problems from counseling individuals to designing interventions that focus on the community. The field has also evolved from being expert driven to developing a partnership with the community in designing and implementing interventions. In solving social problems, the goal is to be faithful to the needs of the community and its members, allowing them to create their own voice and, eventually, sustain their own changes rather than telling them what their problems are and how they should go about addressing them (Rappaport, 1981; Trickett, 1984; Wandersman, 2003). The social agent of change has to act much like a community researcher and have a certain degree of openness to the issues and changes resulting from the interventions (Shweder, 1995; Weisner, 1996). To facilitate an effective relationship between the change agent and a particular community and the creation of a program of change, change agents can learn some lessons from community psychology. These lessons are essential for leading a community to transform itself and include: 1) the change agent locating himself or herself in the community as a passionate facilitator of transformation (Sarason, 2003), 2) defining the problem from the perspective of the community (Rappaport, 1981; Zimmerman, 2000), 3) using multiple methods to measure the problem (Banyard & Miller, 1998; Stewart, 2000), 4) collaboration during implementation (Wandersman, 2003; Trickett & Espino, 2004), and 5) being flexible to change at any time, given the community's evolving needs (Tharp & Gallimore, 1982; Sarason, 2003).

Being an Embedded and Passionate Facilitator

Unlike scientists and engineers who are supposed to be disinterested and dispassionate observers of scientific processes, people interested in social change have to be passionate about the issues they are going to work on and concerned about the community they are working with. Social change agents should ask themselves some tough questions and try to sort out why they are attempting to bring about change in a particular community in the first place. Sarason (2003) emphasizes the need for community researchers to locate themselves and their own motivations in their research. In many disciplines, this may seem counterintuitive to the process, even pointless, but not in community development programs. He argues, "the path of atoms in a cloud chamber do not affect or are not affected by the physicist's personality and thought processes. That is not and cannot be the case in a community intervention" (Sarason, 2003: 209). Because of the nature of social change and its requirement for the change agent to work closely with others, often people from various backgrounds, these agents are not at liberty to ignore their own motivations, expectations and hopes for a project. At some point, they could all affect the way one tries to collaborate, collect data or implement change.

Furthermore, community change efforts require a significant commitment of one's time and energy, due to their complexity and the intricacy of the evaluation process (Tharp & Gallimore, 1982). First and foremost, the change agents need to understand and commit to the time commitment. Locating one's own role in a project or reasons for doing it can be tricky. Sometimes this involves some self-exploration about one's social identity and motivation. If community members and participants are from a different cultural group than the change agent and members of his or her team, it is imperative that the potential effects of this situation be stated and understood before attempting collaboration, as they can create obstacles in the process or unknowingly misrepresent the data altogether (Langhout, 2006; van Uchelen, 2000; Stoecker, 1997). Trust and understanding are built over continued contact with one another, and it is crucial to think about one's own contextual background and how this potentially affects his or her relationship with community members and one's view of the problem. In locating oneself in the community and the problem space, *one has to remain a passionate facilitator, rather than an expert telling the community who should do what, how, and when.*

The embedded facilitator is nevertheless a champion of the project, and needs to find other champions in the community. The concept of champions is borrowed from management literature. It refers to organizational entrepreneurs, or intrapreneurs, who identify new ideas and champion them, or carry them through, from ideation to fruition. Peters and Waterman (1982) presented evidence from American companies that showed that product champions play a significant role in developing and implementing new ideas. For example, in one of their studies they found that in a data set of twenty-four cases, of the fifteen that were successful, fourteen involved a clear product champion. Of the nine failures, just three were champion-led (six either had no champion, or the champion had left early and the project consequently had fallen apart). They also found support for this in Japan. According to these authors, product champions are neither intellectual giants nor dreamers. They are pragmatic and once they identify an idea, their own or someone else's, they doggedly pursue it to the end.

Bhawuk (2001) proposed that in the context of international development projects, donors and recipients needed to identify champions for the projects. He argued that a major reason for the failure of foreign aid projects is the inability of both the donors and the recipients to negotiate who would champion the project. Champions need the support and protection of senior people in the system to carry out activities that may be irregular but necessary to lead the projects to success. A critical aspect of championing a product or project is overseeing it from the beginning to the very end. Since experts are often attached to projects and interventions for a limited time of less than a year to 3 to 5 years, the value of the community or local champion can hardly be overemphasized. It is the local people who are permanently associated with the project and are responsible for sustaining development, and hence it makes sense to choose champions from among them. The experts should play the role of facilitators as mentioned above, and help build the technical skills of the community champions.

A major difference between a product champion and a community project champion would be personality characteristics. James Brian Quinn, a management scholar (quoted in Peters and Waterman, 1982: 206) describes champions as "obnoxious, impatient, egotistic, and perhaps a bit irrational in organizational terms." These qualities seem to characterize managers in individualistic cultures, which are not likely to be universal. The champions of community programs and projects targeting social change would be zealots of change but masters of social and political skills since development requires changes at many levels and only a politically skilled leader can see them through. They must also be sensitive to cross-cultural differences since they have to often interact with people from different cultures. They could also play the role of local informants for the experts who are new to the community. Like the product champion, project champions need not be intellectual giants or creative people to offer unique solutions; they should have the patience and social skills to communicate new ideas and convince people to carry them out. In a nutshell, these champions are sensitive and skilled politicians committed to change.

Defining Problems with the Community

One of the main roles of social change agents is to bridge theory and practice, educating themselves in theories of community development that have been known to work and then identifying and crafting ideas and methods that would fit the need of the particular community they are working with (Julian, 2006; Spoth & Greenberg, 2005; Snow, 2000; Wandersman, 2003). In order to properly apply the theoretical ideas to specific communities in need, the change agents need to collaborate with community members and become partners in the process rather than work as expert leaders (Trickett & Espino, 2004; Pena & Gallagos, 1997). The ultimate goal of collaboration is not to give community members a voice, but to allow the change agents to understand the problem more completely and to emphasize the value that needs to be placed on local knowledge. Trickett and Espino (2004) state, "the importance of valuing local knowledge represents in part a reaction to the kind of expert knowledge social scientists claim as a result of their carefully controlled work. When the concern is making knowledge socially useful, however, the issue of where the knowledge is applied becomes central" (p. 11). The commitment to collaboration and the use of local knowledge means that change agents have to create a space for community members to define the problem for themselves and this calls for anchoring the methodologies in the context of the specific community. It is in this phase that the role of facilitator becomes transparent. An expert tells the community what the problem is and what the solution should be based on his or her expert opinion, whereas a social change agent facilitates the discovery of the problems and their solutions in the context of the strengths and resource constraints of the community.

It is important to examine the concept of superordinate goals here as no community can ever take the first step toward change or transformation if its members are not committed to the goals of the program. When two or more parties are involved, superordinate goals are defined as goals that are not achievable by the efforts of a single party (Sherif & Sherif, 1959). Cooperation of all the parties is a necessary condition for achieving these goals. The goals of community development programs or interventions must necessarily be superordinate to both the community members and the change agents for them to be able to work together to accomplish them. Both the parties must be equally accountable to the achievement of these goals. And this can be achieved by following the above procedure where the expert works with the community and helps the community members develop and implement the program. If the community members are not involved, they will lack the motivation to support the project.

A method of enforcing superordinate goals is to evaluate project members from both sides against the same criteria. This is rarely done in community development programs, since usually the experts who come from outside have more power, and they avoid joint responsibility with the community members. It would be useful to install a 360-degree feedback system, where all stakeholders get to evaluate each other for development rather than for punishing people for making mistakes. A key assumption in carving superordinate goals is that all members have equal access to information. The experts seem to control the information to their personal advantage, and elimination of the information asymmetry between the expert team and the community members is a prerequisite. The key players of the project must have full information about the resources and they should be accountable for its utilization. Setting superordinate goals in this way will engender dual-accountability among the community members and the experts, which is likely to lead to the successful development and implementation of change programs.

Using Multiple Methods and Perspectives

In order to understand a complex community issue, it is best to review it from multiple perspectives, which means that as change agents, we cannot be overly committed to one particular method or viewpoint. It is important to highlight the value of methodological diversity, by understanding that method is only a tool to address the question, and should never automatically define the solution. "The pluralist stance is that traditional scientific approaches (usually quantitative, often experimental) and their alternatives (e.g., qualitative, narrative, post-modern) all have their place and are all to be valued" (Barker & Pistrang, 2005: 202). Because both quantitative and qualitative measures can be used to evaluate community phenomena, it is best if the quantitative measures that have been pre-determined, are informed by the culture and context under study. Social change agents are not immune to pre-conceived notions and cannot be objective; hence *self-exploration is the first step to responsible social change*.

People from different cultural backgrounds view the communities in which they live differently (Dumas, Rollock, Prinz, Hops, & Blechman, 1999; Guarnaccia & Rodriguez, 1996; Cooper & Denner, 1998). While it is the responsibility of social agents to be equipped with scientific tools, it is also their responsibility to be well informed by the participants, regarding their cultural beliefs and viewpoints. Social change agents need to be open to many different methods to accomplish this. One way of extracting this culturally relevant information is via qualitative methods (Stewart, 2000; Banyard & Miller, 1998). Qualitative research methodology allows one to gain insight into the perspectives of the participants, and as a result, the change agent is able to serve the needs of the community as they would themselves define them, not as experts would. Stein and Mankowski (2004) suggest that the process of discovering problems should include asking, witnessing, interpreting and knowing. By following these steps and drawing the participants in the problem definition phase, one is able to reach the research and action goals of the community with them knowing full well what they need and want.

Qualitative measures also help ensure that the cultural differences that exist between researcher and participants are bridged. Grounding research in a mutually created meaning of the problem allows both the change agent and participants to invest in the change that is ahead of them (Wicker, 1987; Hughes & Seidman, 2002). The narratives and meanings invoked through qualitative inquiry allow for a better interpretation of the outcomes of a project and are also useful to inform the other quantitative measures used in the study. Informing quantitative measures using qualitative applications is critical because many times quantitative measures created outside of lived contexts are simply not appropriate (Greenfield, 1997; Marsella, 1998; Stewart, 2000; Gergen, Gulerce, Lock, & Misra, 1996). By conducting focus groups, interviews, narrative, storytelling sessions, or using ethnographic techniques (Weisner, 1996), one is able to get a deeper understanding of the problems, the issues and their probable solutions. Because of the potential change in ideas and goals, it is essential that the social change agents gain an insider's viewpoint of the community when facilitating the development of an effective community program.

Conducting Collaborative Implementation

After obtaining the relevant contextual information and members' perspectives, one needs to collaboratively execute program goals. Often, a compromise between engineering and social science approaches stops the input of the community in defining the problem, and the experts take over the implementation of the solution. It is no surprise that development projects are known to collapse after the experts leave (Bhawuk, 2001), because the community never buys into the implementation of the solution. Much like academic researchers, while working collaboratively and closely with a community project is often difficult for change agents who are working on multiple assignments, it is vital that they stay as connected with the ongoing process as possible (Tharp & Gallimore, 1982; Julian, 2006). Not only will the feedback given by participants and other community members intricately involved in implementation be invaluable to proper execution, it is also important for future implementation efforts (Price & Behrens, 2003; Trickett & Espino, 2004) and theory building. The more social agents share experiences with diverse communities, the better prepared they will be for future work. The involvement in the implementation phase shows that one is passionate about the change process and is willing to stay committed till the end to see the transformation. This highlights the role of the change agent as an insider rather than a dispassionate outside expert, and by sharing responsibility of program implementation, one cultivates trust and further understanding of community members' perspectives (Herr & Anderson, 2005; Schram, 1997).

Furthermore, by remaining involved in the implementation process, one is able to maintain a systems perspective in which sustainable change and future community progress are emphasized. Creating sustainable change is argued to be the goal of empowerment (Rappaport, 1981; Zimmerman, 2000; Fetterman & Wandersman, 2004), which is created by defining the problem in the words and ideas of community members, listening to their stories and life histories, and implementing programs from their standpoint, as discussed above. By empowering, one gives the words and ideas of the community the agency to change.

An important component of whether a community can sustain needed changes, is whether members feel a sense of collective efficacy and hope that their situation can change for the better (Foster-Fishman, Cantillon, Pierce, & Van Egeren, 2007). By empowering their words and stories, obtained through careful qualitative methods and culturally anchored quantitative ones, there is a likelihood that members will feel as though they have been heard and their ideas are central to the ultimate goals of the community program. Following this protocol will help to ensure that members have a stake in the success of the endeavor, which creates a more lasting ideal (Zimmerman, 2000; Stewart, 2000). With community voices elevated, the next procedural step would be to illuminate the goal of system change. A good marriage between a community and the social systems in place to serve it is one of "competent individuals involved in responsive social systems" (Kelly, Ryan, Altman, & Stelzner, 2000). By illuminating the ideas of the community through informed methods and empowerment, one allows to foster community members' competency and stake in their community's future. This helps create responsive social systems.

Participatory action research fosters a high level of commitment to change from community members and social change agents. When community members become active units of change, the goals of the intervention are that much more likely to be met. Instead of the traditional social engineering approach, which is expert-led, where change agents "lead the pack" toward the goal with little community collaboration, change agents need to be steadfast in their responsibility as cooperative change agents supportive of direct citizen participation (Wandersman & Florin, 2000; Foster-Fishman et al., 2007). By increasing the responsibility and feeling of control over the particular program, and by implementing participatory interventions, social change agents increase the buy in from the members' point of view and, as a result, are likely to increase the efficacy of the program (Wandersman, 2003; Schorr, 1997). Wandersman and Florin (2000) noted that "participation increases feelings of helpfulness and responsibility and decreases feelings of alienation and anonymity" (p. 247). When this participatory action component is combined with culturally anchored research and empowerment, the renewed confidence and voice of community members is likely to enable them to challenge the status quo and work to change the social systems which will then be able to help even more people. Challenging the efficacy of poor social systems by loaning needed communities an active and participating voice is certainly a valid procedural step in this model of effective community change.

The legacy of participatory research lies with Paulo Freire (1970). The basis of Freire's ideas is a critique of the dominant culture's presumed expertise regarding the social ills of those in other, less dominant, groups (Herr & Anderson, 2005). While Freire's writings were dominated by the liberation movement happening in Latin America at the time, the tenets of action research, at their core, employ qualitative methodology in order to empower citizens. One of the general characteristics of action research is for the change agent and the community to create new and useful knowledge that can lead to social transformation. The goal of the social change agent is to extract meaning from participants, not only to work to create the best possible intervention that fits their needs, but also to motivate and empower them to focus on the systemic changes needed. Thus, social change not only requires superordinate goals but superordinate implementation procedures.

Being a Flexible Change Agent

Finally, because implementing a community program involves so many stakeholders, it is important to be flexible all the time, from problem definition to solution implementation, to accommodate a variety of opinions as well as the changing circumstances of the community involved. Sarason (2003) notes, "an intervention is an exercise in diplomacy, which is the art of compromise based on the recognition that each of the parties cannot have it all its way" (p. 211). Sarason (2003) points out that if there are a number of people involved in implementing a community program, then all those with opposing ideas from that of the change agent must still be considered and understood. Along with creating the program based on the community's definition, there are bound to be members who disagree and have the right to be heard. Retaining one's own flexibility in construction and evaluation is fundamental to make sure that the community needs are met over the length of one's time working with them (Tharp & Gallimore, 1982).

It is perhaps being inflexible to suggest a specific protocol for community development work. Since each community is different and requires a different approach to solutions, points two, three and four discussed above will be influenced by the fifth point—flexibility will be needed in problem and solution definition as well as implementation of the solution. And no project of social change can make any progress if the facilitator is not passionate about the issues, and deeply concerned about the wellbeing of the community, which was the first point we discussed. Empowerment, participatory action and the unique and culturally anchored mixed methods that promote their inception are ways to ensure that we are properly representing community members and staying flexible to their desires. Working as community scientists forces us to juggle our lives as objective change agents and passionate and humanistic members of a community in which we locate ourselves, and while this is, at times, a difficult task, it creates an opportunity to be a living bridge between people and ideas. Social change agents cannot afford to be expert engineers talking down to people and communities, and they have to evolve into becoming passionate community members who can lead the people to identify their own needs and ways to meet those needs, for the best leader is one who is not missed by the empowered community.

Bhawuk (2001) presented a template for effective international development project negotiation that seems to be equally applicable for developing any community project. He used the metaphor of a suspension bridge, and suggested that superordinate goals and champions constituted the two pillars necessary to construct the bridge to build a development project. We saw in the above discussion that these two constructs are embedded in the five-step process. Thus, developing superordinate goals and finding and nurturing community champions would be two other criteria that would help go beyond social engineering toward community transformation. It should be noted that principles of collaborative participation and empowerment are also at the core of the model.

Three Cases Supporting the Model: The Community and the Change Agent

Exemplar programs that have used the above procedure or some variation of the ideas included in them to help transform a community are available all over the world. Mohatt, Hazel, Allen, Stachelrodt, Hensel, and Fath (2004) reported a community psychology intervention program in Alaska that successfully used most of the above principles. In the following, three case studies are presented to demonstrate the applicability of the model to community transformation. These three organizations and the change agents who were instrumental in starting them are Dr. Kurien of Amul in India, Dr. Yunus of Grameen Bank in Bangladesh and Father Don Jose Maria of Mondragon in Spain. These cases are from two continents—Asia and Europe, are spread over three industries—agriculture, banking, and manufacturing, and cover farmers, micro-entrepreneurs, and people with technical skills, thus allowing for the potential implementation of the protocols across cultures and industries.

Amul and Dr. Kurien

Amul (Anand Milk-producers Union Limited) is a name known to all Indians, which is an exemplar cooperative demonstrating the value of community participation in change management. Amul was organized as a cooperative by Mr. Tribhuvan Das Patel in 1946, and he invited Dr. Kurien to lead the cooperative in 1949. Amul has more than 2.6 million members who are involved in milk production in Gujarat and its annual turnover is over US$1 billion. Amul started the "white revolution," and has led India to be the largest milk producer in the world, collecting 6.5 million liters of milk every day from some 12,800 village cooperative societies in Gujarat. It has 13 district dairy unions where milk is stored and processed and milk products are manufactured.[3]

Being an Embedded and Passionate Facilitator

Dr. Kurien is known as the "father of the White Revolution" and the "Milkman of India." In 1946, milk distribution in Bombay was monopolized by Polsons Dairy, a British company. Polsons had a monopoly and it forced dairy farmers to sell their milk, a perishable product, at a price that was extremely low compared to the final selling price. Hence, Amul was born as a protest to this practice. Dr. Kurien later described that "Amul's birth was thus a harbinger of the economic independence of our farmer brethren. Amul's mission was the development of farmers, nutrition to the nation, and heart in heart, the real development of India" (Kurien, 2001). The choice of the word "farmer brethren" is not merely cultural and is indicative of the community that Amul has been from the start.

Dr. Kurien described his humble beginnings in the town of Anand in 1949, "a dusty small town of 10,000 people," which was anything but pleasant. He narrated his story in an interview (Roger, 2007): "I was compelled to come here because my education was paid for by the government, then British government. Therefore, my obligation to the British government was transferred to the Indian government. So I was sent to Anand. People were not modern. Road, communication, everything was bad. There was no bathroom. Three corrugated sheets made my bathroom. That is how I started." He could not find a decent rental place because he was "Christian, and outsider, and above all a bachelor." So he lived in the garage of the dairy plant. He had to construct his own lavatory. He lived in the same modest condition as that of the community he represented. But he stayed back because of the warmth he received from the farmers. And he states, *"It is here that I found myself, and I am glad that I stayed here."* The world-renowned social entrepreneur who created a world-class organization, and led India to become the largest milk producer in the world in his senior year claims to have found himself in Anand. This shows the power of mutual transformation. *When a change agent stays committed to the community, the transformation is bidirectional, and the change agent transforms with the community.*

It is plausible that the farmers could relate to him as he was no different from them in his living conditions despite his elitist education. In the interview, he further noted: "I could have gotten a high paying salary in a city. But I could not have received the warmth, affection, and the love of the people that I worked with, and people for whom I worked in those environments."

Defining Problems with the Community

Some argued that Amul's "system which involves participation of people on such a large magnitude does not confine itself to an isolated sector. The ripples of its turbulence affect other areas of the society as well."[4] This is certainly true and Amul has impacted the transformation of the Indian communities in many ways. For example, Indians are drinking three to four times more milk than they did four decades ago. This has immeasurable impact on the health, nutrition, and life expectancy of the Indian population.[5]

However, the greatest impacts were experienced by the villages that participated in the Amul program. Dr. Kurien described in an interview how Amul has effected social change at the village level in multiple ways (Roger, 2007). For example, people stand in line to deliver their milk without trying to take short cuts. They do not complain if they have to stand behind an untouchable in the line. Hence, the cooperative has thus given a deathblow to the caste system in its own way. Another example is that Amul provided women an opportunity to have a voice in the "home economy" as they are the major participants in the program.[6]

In an interview, Dr. Kurien had remarked that there was no way to replicate Tribhuvan Das Patel, the farmer leader who started the cooperative in 1946. However managers like himself could be developed through education and training. Thus, recognizing the importance of both managerial professionalism and the specific challenges of rural management, Dr. Kurien advocated the creation of the Institute of Rural Management, Anand (IRMA) to train young aspirants to become professional rural managers. One of its main missions is to "educate a new breed of professional rural managers having appropriate values and ethos to help rural organizations and institutions in professionalising their management and empower rural people through self-sustaining processes."[7] Community building requires allowing leaders to emerge from the community who understand the needs of the community. This is not to downplay the role of the expert, which clearly in this partnership is represented by Dr. Kurien, a mechanical engineer by training.

Using Multiple Methods and Perspectives

Amul has many times gone against the expert wisdom, and yet succeeded because of its willingness to try multiple methods and perspectives. Dr. Kurien related that, "in the early days of Kaira Union there was no dearth of cynics. Could natives handle sophisticated dairy equipment? Could Western-style milk products be processed from buffalo milk? Could a humble farmers' cooperative market butter and cheese to sophisticated urban consumers? The Amul team—farmers and professionals—confounded the cynics by processing a variety of high-grade dairy products, several of them for the first time from buffalo milk, and marketing them nationally against tough competition" (Kurien, 2001).

Amul has not only installed more than 4000 Automatic Milk Collection System Units (AMCUS)[8] at the level of Village Societies, but also proved the cynics wrong by successfully producing powder milk from buffalo milk. In fact, in 2005, the total annual production of milk product from buffalo milk was higher than cow's milk and its contribution in terms of financial value was twice as large as the contribution of cow's milk.[9] This is despite the fact that many years ago the expert from New Zealand advised Dr. Kurien that they should not try to make powder milk from buffalo milk because it would not work. Amul went ahead and did it and now makes more powder milk than New Zealand and is the largest producer of milk in the world (Roger, 2007).

Conducting Collaborative Implementation

Dr. Kurien is a true believer of collaborative implementation. Between 1970 and 1996, he established a national dairy program popularly known as "Operation Flood." Through this program, he made possible for dairy farmers to own and operate milk production in urban areas of India. Not only was this program a huge success that helped dairy farmers to improve their financial well-being, it also contributed to the improvement of the infrastructure of rural areas that were involved in it.

> Endowed with decision-making capacities, some leaders in cooperative-member communities have built facilities like libraries and healthcare centers with their profits. The success of the Operation Flood management model led to its application to other commodities. Fruits and vegetables are now produced and marketed through a cooperative system involving a network of over 250 farmer-owned retail stores in Delhi.[10]

When Dr. Kurien was awarded the World Food Prize in 1989, Dr. Norman Borlaugh, the founder of World Food Prize honored Dr. Kurien as "one of the world's great agricultural leaders of this century" for his dedication "to streamlining management and distribution strategies with the skills and knowledge of rural and small-scale producers."[11] Hence, what Dr. Kurien achieved in Amul was not a scientific discovery but rather a result of working closely with the dairy farmers.

Being a Flexible Change Agent

As proposed in the model above, the experts have to be passionately involved, yet be flexible in allowing the community to take its own course, sometimes painfully slowly, to decide how they would like to develop. *Community transformation is a slow process and cannot be managed like profit-oriented organizations that are hungry for quarterly growth.* Dr. Kurien faced an uphill task in the initial years of Amul, but he was flexible to seek remedies. For example, the rapid growth of the Amul movement resulted in overcapacity of milk in winter months when the production of milk was on average 2.5 times higher than other months. Farmers again were forced to sell at lower rate to middlemen. Dr. Kurien among others sought the assistance from the Central Food Technological Research Institute (CFTRI) in Mysore to produce baby food and cheese from buffalo milk. This was the world's first commercial cheese and baby food production from buffalo milk.[12] His skillful handling of challenges and moving forward earned him the praise of Prime Minister Lal Bahadur Shastri on many occasions in the sixties, and he commended his "extraordinary and dynamic leadership."[13]

Dr. Kurien's flexibility to learn and adapt is still evident after more than five decades. He remarked, "In Amul we have a commitment to total quality. But, occasionally, we may make a mistake—or, our customer may think we've made a mistake, and the customer, as they say, is always right. That is why, for Amul, every customer complaint must be heard—not just listened to. And, every customer complaint must be rectified to the extent humanly possible" (Kurien, 2001).

In an interview, his daughter, Nirmala Kurien said, "Nelson Mandela wrote in his autobiography that if a person has a very good mind, and a very good heart, the person is unsurpassable. And my father has both those qualities. (…) He is very passionate about India and what India can do, and what India will be." When asked, as a man who has become a legend in his lifetime, how he would like to be remembered in history he remarked, "As a man who tried his best to improve the living standard of the milk producers of India. Most of whom are small farmers, marginal farmers, and landless laborers. So I was able to help in my own way to raise the living standard of the poor. That is how I would like to be remembered." Again, we find that the identity of the change agent has merged with the community.

Dr. Muhammad Yunus & Grameen Bank

Dr. Muhammad Yunus is an exemplar social entrepreneur, and has eloquently argued that access to capital is a fundamental human right (Hertz-Bunzl, 2006). He has created a community where this fundamental right is guaranteed by members of the community rather than by the goodness of heart of donor countries or the World Bank. He created the Grameen Bank in 1976, which literally means a village bank. Grameen Bank has since reached nearly 7.34 million borrowers of which 97 percent are women and 50 percent of them have successfully crossed the poverty line (Ford, 2007).[14] Counting an average of 6 people in a family, one could say that Grameen Bank has directly benefited more than 42 million people in Bangladesh, accounting for nearly a third of the population of the country.

When Dr. Muhammad Yunus, winner of Nobel Peace Prize for 2006, started in 1976 a Bank Project with a unique service for the poor of Bangladesh, loan without collateral, he clearly understood the need of the community. People who came to borrow from him had nothing of value to give in collateral, and therefore could never qualify to get a loan from a traditional bank. In 1983 he created the Grameen Bank, which stands for a new idea, micro-credit, and serves more than five million people, who have started small businesses or services. The bank is owned by the borrowers, and they own 94 percent of the equity of the bank (6 percent is government owned). Today, Grameen Bank has transformed into a family of services including Grameen Trust, Grameen Fund, Grameen Fisheries, Grameen Telecom, Grameen Shakti, Grameen Phone, Grameen Education, Grameen Knitwear, Grameen Solutions, and so forth.

The success of Grameen Bank has caught worldwide attention and Grameen's model has been replicated in various countries (Ferguson, 2007). The Good Faith Fund in the United States and the Amanah Ikhtiar in Malaysia (Ikhtiar Trust of Malaysia) are some examples (Auwal, 1994). An analysis of Dr. Yunus' effort shows that rather than solving the credit problem in a banking sense as a technical expert, he led the community, intuitively and naturally adopting the five principles of social change stated above, to a transformation that has become an exemplar in its own right. In what follows we discuss how the five principles of community transformation seem to get validation from the work of Dr. Yunus.

Being an Embedded and Passionate Facilitator

Yunus is the "the father of microcredit," which refers to giving small loans to people who would never qualify for it under the most generous of procedures adopted by a commercial or development bank. Following this unique idea, which emerged out of necessity in Bangladesh, he has become today one of world's best known social entrepreneurs (Hertz-Bunzl, 2006). Yunus has created a system where every applicant acts as collateral for each other, every applicant automatically becomes a stockholder of the bank and their voices are represented in Grameen's decision-making processes. This system supports the emergence of local champions, and is a part of Grameen's "standard operating procedure." Applicants are expected to be responsible and accountable for their well-being, the well-being of their peers and the well-being of Grameen Bank.

By creating champions, Grameen becomes a facilitator of change rather than an organization that carries out the change by itself. Grameen Bank does not allow the poor to become dependent on it, but acts as a catalyst to help them overcome poverty on their own. Hence, Grameen Bank does not operate as a welfare organization but as a truly profit seeking organization. The interest rate charged by Grameen Bank is 5 percent higher than what would be offered by the commercial banks in Bangladesh (Auwal, 1996). This is quite paradoxical but Grameen's rationale is that a higher interest rate is needed to effectively serve the population that has been neglected by the commercial banks. For example, Grameen believes in visiting potential loan applicants rather than waiting for them to come to the bank. This seems like an inefficient and expensive process but has counter-intuitively been effective in addressing poverty. Giving loans to beggars to sell popular consumer items is another example that many commercial or government banks in any part of the world could not even think about. However, by providing a loan of USD $10, Grameen has successfully transformed many beggars from being "losers" and a burden on the society to becoming self-supporting "champions" controlling their own well-being.

Grameen has demonstrated that it is possible to remain a facilitator even in emergency situations that often call for leaders to take control and direct rather than allow the community to take charge and manage the crisis. In 1998, when Bangladesh had one of the worst floods that destroyed the economic base of a large population; Grameen took a massive initiative in providing new loans, deferring old loans and providing basic relief goods for people impacted by the flood. This was consistent with Grameen's policy of charging fee for the services they provide for victims of natural disasters and calamities. Though all these efforts were done in a truly humane spirit, all of them were accounted for accurately on a commercial basis and none was written off. Yunus remarked that "if you give it free, like the oral saline solutions, they will waste it away," and he responded to the critics by arguing that "Grameen Bank is the people's bank, it has to protect its interest for the people who own it, to the point of charging those same people for relief services extended to them." He further argued "that is what makes the bank stable and prosperous: The people who own it."[15]

Defining Problems with the Community

Grameen approached poverty reduction from an "action research" perspective (Auwal, 1996), which emphasizes understanding the environment being studied. Hence, Grameen's approach embeds it in the community. New employees, for example, are required to stay in a village for a year and are asked to actively participate and learn reflectively from Grameen's approach to poverty issues and their alleviation, which are often emergent in nature (Auwal, 1996). Grameen recognizes that poverty is not necessarily a result of illiteracy but rather a result of inaccessibility to capital. Hence, the thrust of Grameen's approach rests on the belief that the poor have adequate 'survival skills' to compete in the capitalistic world provided they can obtain capital (Yunus, 1987). Nobel Laureate Amartya Sen has noted that Grameen's approach is as much in line with Adam Smith and Peter Bauer's notion that the poor can compete in a capitalistic world provided they have adequate institutional support (Sen, 2005).

Realizing that poverty is not an isolated issue from the overall social development, Grameen worked collaboratively with the poor in developing what is popularly known as "*Sixteen Decisions*." Among others, these "sixteen decisions" are hoped to address the need for the poor to raise their social, cultural and political consciousness. For example, the practice of dowry that is prevalent in Bangladesh is not only morally wrong but also perpetuates the cycle of debt and poverty. Hence, one of the sixteen decisions requires every borrower to commit against the dowry system by stating—"We shall not take any dowry at our sons' weddings; neither shall we give any dowry at our daughters wedding. We shall keep our centre free from the curse of dowry. We shall not practice child marriage."[16] The "sixteen decisions" are in a way a binding contractual agreement between the bank and the poor, which is monitored by the community not the bank.

Using Multiple Methods and Perspectives

Dr. Yunus and Grameen champion multiple methods and perspectives both to understand the problems of the poor as well as to generate solutions for the poor. These approaches include the need for Grameen to be in the field. Scholars have noted that Grameen is successful in reducing poverty in Bangladesh because it combines both informational and communicational approaches. Integrating these two approaches, Grameen has leveraged the creativity of its employees as well as clients (Auwal, 1996).

Yunus tells how he started Grameen, which illustrates the value of using multiple methods:

> "It made me realize that, whatever I had learned, whatever I was teaching, was all make-believe; it had no meaning for people's lives. I traveled around the village and talked with its people. Soon, all my academic arrogance disappeared. I realized that, as an academic, I wasn't really solving global problems; I wasn't even solving national problems. I decided to abandon my bird's eye view of the world, which allowed me to look at problems from above, from my ivory tower in the sky. I assumed, instead, the worm's-eye view and tried to probe whatever came right in front of me—smelling it, touching it, seeing if I could do something to improve it. Trying to involve myself in whatever capacity I could, I learned many things in my travels" (Yunus, 1997: 1).

What Dr. Yunus calls a worm's-eye view can be viewed as a metaphor for an alternate methodology. After all he was open to accept that his bird's-eye methodology was of no help, and was willing to adopt a new approach where he was embedded in the community he was trying to help, and he could smell and touch the change he was trying to make.

Conducting Collaborative Implementation

All the borrowers automatically gain the status of Grameen's stockholder. They are represented in the decision-making processes at the board level. By involving the poor and often uneducated in the decision-making process, Grameen has continued to collaborate and also involve the community members in the implementation of all their programs. It demonstrates that the poor also have wisdom and competence to take care of their needs. All they need is the opportunity, and Dr. Yunus and Grameen have provided such opportunities in an environment of support (Auwal, 1996).

Grameen's solidarity group approach (Dana & France, 1996) in where they ask each borrower to form a group of five of similar needs and backgrounds and delegate the repayment responsibility to the group is not only culturally appropriate but also provides the borrowers an opportunity to collaborate with each other. Hence, not only Grameen collaborates with the borrowers to generate solutions but also encourages the borrowers themselves to collaborate with each other for their individual and collective success (Bhattacharyya, 2004), leading to the transformation of the community. These collaborative engagements also raise the level of the borrower's engagement, commitment and pride in Grameen, which is reflected in many ways. For example, when Dr. Yunus was awarded the Nobel Peace Prize in 2006, nine women from the villages accompanied him to Oslo and millions of borrowers were glued to their TV sets watching the award ceremony. There was a feeling of euphoria among the 7 million borrowers-cum-owners.[17]

Being a Flexible Change Agent

Grameen's approach of decentralization, flat-bottomed structure and collaboration with the borrowers allows the organization to be responsive to the changing and emergent needs of the poor. It also allows for assessing and anticipating such needs (Auwal, 1996). The quote below is a testimony for the change agent to be flexible, and Dr. Yunus states in his own words how he took one step at a time, without doing much sophisticated analysis that is known to lead experts to paralysis. Dr. Yunus and Grameen illustrate the value of allowing the solutions to emerge from the context naturally rather than imposing prefabricated solutions:

> "Let us not expect that a social business enterprise will come up, from its very birth, with all the answers to a social problem. Most likely, it will proceed in steps. Each step may lead to the next level of achievement. Grameen Bank is a good example in this regard. In creating Grameen Bank I never had a blue-print to follow. I moved one step at a time, always thinking this step will be my last step. But it was not. That one step led me to another step, a step which looked so interesting that it was difficult to walk away from. I faced this situation at every turn."[18]

Don Jose Maria and Mondragon

Based in Basque Country, Spain, Mondragon Corporación Cooperativa (MCC) is one of the world's best known worker cooperatives (Thomas & Logan, 1982). The success of Mondragon Cooperative Corporation (Whyte & Whyte, 1991; Morrison, 1991; Thomas & Logan, 1982) is an example of building a community through economic activities based on democratic principles of participation and one person one vote, unlike the stockholder system where one person can have more say than hundreds of people because they have invested more money.

Starting from scratch, five young engineers founded Mondragon in 1956 with inspiration and guidance from a priest, Don Jose Maria. Today, it has expanded its operation to over 10 countries, employs more than 80,000 workers, has wide ranging portfolios in manufacturing, finance, distribution, research and training, and is twice as profitable as any other Spanish corporations.[19] Don Jose Maria never occupied any executive position, but still played a key role in the success of Mondragon:

> "With some projects it is almost impossible to identify the single spark which lit the flame, often because such a spark simply does not exist. In this case, however, the writing is on the wall and even those of us who were not lucky enough to have played an active role in the group's initial development can clearly see that Father José María was indeed the fundamental driving force behind the whole experience."[20]

Being an Embedded and Passionate Facilitator

Don Jose Maria never had a plan to be at Mondragon and tried to persuade his superior to allow him to pursue a graduate degree in sociology at the University of Leuven in Belgium. His superior, however, refused and he ended up in Mondragon, a town of 8000 people known for its poor economic condition (Russell & Rus, 1991). He took the new responsibility seriously, and considered himself as a teacher rather than a preacher. He often went beyond his normal pastoral duties to engage the local community in the discussion of local issues (Russell & Rus, 1991).

The desperate economic and social situation in the community inspired him to build Mondragon economically and socially. A strong advocate of "knowledge is power and that people must raise themselves by their own efforts,"[21] he established a technical school as a vehicle to overcome Mondragon's social and economic challenges (Miller, 1996). In 1956, when five young engineers who were his former students approached him for the possibility of setting up Ulgor (a kerosene stove factory) in Mondragon to help build Mondragon's economy, he was eager to help. During the next 20 years, he was instrumental in ensuring that the economic activities were embedded in the principle of human dignity, democratic cooperation, solidarity and the importance of knowledge (Miller, 1996). He was often criticized for mixing religion and worldliness, and his response was—"If the Gospel does not apply to the economy, then to what does it apply" (Vincec & MacLeod, 1996: 23).

Don Jose Maria's passion for Mondragon sometimes put him in difficult situations. For example, realizing that Mondragon desperately needed a financial arm and failing to get the attention of his associates, he forged two signatures of the founders of Ulgor to obtain approval from the government to establish a co-operative bank (Caja). His credibility did not diminish because his intention was only to help the cooperative, and so two of the founders of Ulgor did lead the bank, which is perhaps the most important arm of the Mondragon enterprise today (Russell & Rus, 1991).

Defining Problems with the Community

When Don Jose Maria first arrived in Mondragon, he noticed that emigration of youths due to the lack of employment opportunities in Mondragon was a serious problem. He felt the future of the Mondragon community would be bleak if the youths left the community, and so working closely with blue-collar youths and their parents (Russell & Rus, 1991), he established a technical school to help Mondragon's youth to acquire necessary skills to be more employable in Mondragon.[22] His approach of working closely with the community and defining problems with them was also evident in Mondragon's (MCC) core principles. For example, every cooperative needs to operate on the basis of participative management and has social responsibility (Clark, 2004). The cooperatives are open to self-criticism, hold open discussions, and work closely with the political and economic realities of Spain (Russell & Rus, 1991). Don Jose Maria's approach of defining and solving problems with the community is also evidenced by the fact that he leveraged the cultural aspect of the Mondragon community. As the Mondragon culture valued solidarity and teamwork, he successfully transformed those practices in Mondragon's (MCC) daily operations (Whyte &Whyte, 1991).

Using Multiple Methods and Perspectives

Don Jose Maria is a keen believer of "testing ideas in the real world" (Vincec & MacLeod, 1996: 23). His belief was deeply influenced by the motto "See, judge and act" of the Young Christian Worker's movement in which he was actively involved (Vincec & MacLeod, 1996: 23). Though he valued and taught the Church teachings, he found them to be too abstract for direct application in real life. His deep conviction that ideas need to be tested led him to experiment with some of the dominant thinking and assumptions of economic and social order. Specifically, he challenged Emmanuel Mounier's assumption that humanistic cooperative business could not succeed in capitalist markets. Through Mondragon's (MCC) success of infusing capitalist managerial technique in humanistic cooperatives, he proved Mounier to be wrong (Vincec & MacLeod, 1996; Guillé, 2001).

He believed that ideas can be refined by listening to people with openness, debating on them, and trying them out. These values are deeply embedded in Mondragon's culture. In fact, Mondragon has a dedicated organization (EZAI) with the sole purpose of promoting research into public policies (Mondragon Corporation Cooperativa, 2006). His emphasis on practicality and striving for multiple perspectives is also evident in Mondragon's organizational structure and how decisions are made. It uses a Governing Council and a Social Council; the first is responsible for managing tasks and profitability and the second for concerns about people (Heller, Pusic, Strauss & Wilpert, 1998). The Social Council functions much like the employees' union and the Governing Council like management, and differences that cannot be negotiated are resolved at the annual general meetings through the democratic principle of voting on unresolved issues (Whyte & Whyte, 1991).

Conducting Collaborative Implementation

Mondragon puts social development at the core of economic development goals (Shipp, 1996), and collaboration with the community is its standard operating procedure. It is noted in Don Jose Maria's written reflection—"The self-managed society will be that in which all of us, with our education and willingness to participate, are able to realize accomplishments [Reflection, 20:57]" (Herrera, 2004). This is also clearly reflected in Mondragon's notion of "joint self-employment" (Lutz, 1993). What this means is that every employee is a co-owner and responsible for collective goals. This is true not only within Mondragon's cooperatives and between Mondragon's cooperatives but also between Mondragon and the community they serve. Shipp (1996) argued that the ability to foster cooperative entrepreneurship and striving for collective economic advancements are some of the unique characteristics that contributed to Mondragon's success. Community psychologists have also argued that Mondragon (MCC) is an exemplar of how "empowerment outcomes at the community level of analysis are expressed in multiple empowered organizations within a community and collaboration across multiple sectors within a community. Ideally, communities of this sort provide multiple opportunities for their citizens to participate and shape community life" (Speer & Hughey, 1995).

Being a Flexible Change Agent

Though Don Jose Maria failed many times in his effort of transforming Mondragon's community, he was persistent and flexible enough to accommodate different strategies. For example, when he failed to convince the "Unión Cerrajera to open its apprentice school, he organized the community to create the Escuela Politécnica Profesional for training young boys in industrial skills" (Russell & Rus, 1991: 66). Don Jose Maria's flexibility is well reflected in how Mondragon adapted its business strategy to the changing external environment. It also constantly invested in technology and worker skills (Guillé, 2001). For example, changing the ratio of flat wage structures from 1:3 to 1:6 (the highest salary is not more than 6 times of the lowest salary) is an example where Mondragon (MCC) adapted to the capitalist ideology (Guillé, 2001). Another example is when Mondragon formulated an option for non-members who work for Mondragon's publicly listed companies to participate in the ownership and management of their organization.[23]

Mondragon's ability to be a flexible change agent is well described by Kanter (1972: 133) and Stryjan (1989: 43). Kanter noted that Mondragon "developed their communities by stages.... Members often made choices at each step of the way. ... The full organization grew out of a series of smaller steps and built on existing commitment as the base for generating more commitment." This was echoed later by Stryjan who noted that Mondragon has "in fact, evolved gradually, nearly accidentally, in a succession of organizational choices and changes and not as a realization of a preconceived plan."

Discussion

We started the chapter by discussing the paradigm used by engineers in solving problems and suggested that this paradigm was limited in solving human problems by virtue of being mechanistic and thus leading problem solving through legislation and procedural changes, often coming from experts who are unaware of the issues the community faces. Critics of the social engineering approach to institution building have faulted this approach from the outset as static, a priori, and unoriented (Siffin, 1972; Ganesh, 1980). In contrast to the social engineering approach, the protocol presented above recommends empowering the community to define and solve the problems in their own terms, using their own language and community expertise. The three cases we presented support the idea that when people in a community come together, they transform not only their own lives but also those of the many other community members they interact with, and their unsophisticated ideas far bedazzle the solutions proposed by experts.

It should be noted that Grameen and Mondragon have both instituted collaborative participation as their standard operating procedure. Though it is not clear if participation is a part of Amul's operating procedure, their operation clearly demonstrates collaborative participation. It should also be noted that in all these organizations, the founders acted in more ways than the traditional sense of being facilitators, and were deeply involved in making many major decisions with the community. They demonstrated that collaborative participation includes the social change agent and it is expected that the change agent will work in the community as a community member, rather than a distant observer—a social change agent is not a catalyst that leaves the chemical reaction unchanged but rather is also transformed in the process.

This chapter has implications for institution building in that it invites researchers and practitioners to shift the focus from one approach to another. Instead of "planning, structuring and guidance of new or reconstructed organizations" (Esman, 1972: 22), which is driven by the creation of an organization as a tool to bring about change in the society, it recommends allowing people in the community to define their need in their own words and to find their own solutions, however unsophisticated it may look to the trained experts. To think that organizations that would "embody changes in value, functions, physical and/or social technologies" (Esman, 1972: 22) can be created in a planned manner might have been in the zeitgeist of the 1960s and 1970s, but the dismal performance of development projects all over the world in the developing countries clearly speaks to the failure of this top-down scientific management (Taylor, 1911) approach to institution building. New organizations of change cannot be fashioned by social architects and engineers in their design office unless the community is involved in defining, designing, and implementing the changes on their own terms, at their own pace. To quote Kaplan (2002), "In order to be involved in, or be a guide to, the transforming of the present into a future state—the essence of our work—one must have the skill to do this in such a way that the object with which one is working is not violated, but is transformed according to its own laws" (p. 140). Thus, a shift can be noted in the community of experts that support Community-Driven Development or CDD that is founded on the basic principle of empowerment (Esman, 2003), which is closer to the protocol presented above.

An important lesson from this analysis pertains to the operation of NGOs (non-governmental organizations) and INGOs (international NGOs) in capacity building and helping the needy people. These organizations often work in the social engineering paradigm where they raise funds from donors who are committed to serving a cause, be it saving the children, saving the environment, alleviating poverty, or helping fight injustice, and hire experts to go and carry out projects in different parts of the world. NGOs and INGOs work as brokers who match the donor and the receiver, and never build a community that can help itself in perpetuity. If these organizations were to leave the social engineering paradigm and adopt community transformation, they would be able to help the global community better. It is plausible that if a community developed in one part of the world that learned to save their children, then this community could replicate itself in other parts of the world, and permanently resolve the problems facing orphans and other needy children. Clearly, social engineering has limited success in bringing about social change, and there is a need to move toward the community transformation paradigm.

All three cases highlight the importance of finding culturally appropriate solutions, in defining the needs of the community, measuring the variables of interest, exploring alternative solutions, and implementing the chosen solution. They also demonstrate that social transformation follows a meandering path rather than an engineered blueprint. None of these communities started with an equivalent of a Marshall Plan (European Recovery Plan), but evolved one step at a time, often fortuitously and what they have been able to achieve competes with some of the largest planned interventions including the Marshall Plan. Mondragon has achieved much more than what the Marshall Plan was able to achieve in Europe. Similarly, the Bhakra Nangal Dam (built over 25 years from 1948 to 1963), which was praised by Late Pandit Jawahar Lal Nehru as the "New Temple of Resurgent India," does not compete with Amul and its contributions to the nation. And no effort in Bangladesh comes closer to the impact that Grameen has been able to effect through transformation of the lives of poor people.

The three exemplars that we presented in the chapter stand for three basic human rights—Right to Livelihood (Amul), Right to Credit (Grameen) and Right to Employment (Mondragon). Basic human rights like these are the need of the people at the Bottom of the Pyramid (BoP) (term coined by Prahalad & Hart 1998, to capture the vast worldwide population that earns less than US$ 2.00 per day), and the way to serve them is not necessarily to sell a product to them, but to partner with them and to empower them, so that they can become experts in their own small local area and lead people around them to become self-sufficient. It is about social transformation, and economic development is an emergent outcome of such a transformation. Other such human rights could include right to live, right to health, women's right to live as they choose to, right to good environment, right to ethnic identity, right to live without humiliation, right to practice one's religion, and so forth, and it seems that only social transformation can lead people to secure these rights.

The three case studies show that the five principles identified in this chapter—the need for the change agent to be embedded in the community as a passionate facilitator of transformation, defining the problem from the perspective of the community, using multiple methods to measure the problem, collaborating during implementation and being flexible to change at anytime—are found to be useful across cultures. These principles are likely to be true for community transformation in developed countries as well, especially where community building rather than developing an organization for some stated objective, for profit or otherwise, is the goal. However, these cases provide evidence that, at the operational level, culture has a significant role to play and these principles will need to be adapted to the local ethos to be relevant. In the case of AMUL, we find that despite the norm rather than attitude driven collectivist culture in India (Triandis & Bhawuk, 1997), AMUL was able to drop the trappings of the caste system and cultivate the queue system at the milk collection centers where people stood in line without paying attention to whether the person next to him or her was an untouchable. Grameen has adopted the group approach (Dana & France, 1996) where each borrower is asked to form a team of five people of similar needs and backgrounds. This approach delegates the moral responsibility of repayment[24] to the group and by so doing not only takes advantage of the in-group cohesion found in collectivist cultures (Triandis & Bhawuk, 1997) but also provides the borrowers an opportunity to both expand the constitution of their in-group and work in the comfort of their in-group members. Finally, Don Jose Maria successfully adopted solidarity and teamwork practices that are valued in Spain's culture in Mondragon's (MCC) daily operations (Whyte &Whyte, 1991) by allowing the voice of workers in the shaping of the policy of the organization. Thus, it is clear that in all three cases, collectivist practices of valuing in-group have been used in unique ways to transform the community to cultivate new values that serve both the community and the individuals. It seems that a culturally appropriate solution can always be found to even challenge values that need to be changed (e.g., the caste system in India) for development initiatives to be sustainable (Dale, Ling & Newman, 2008).

Across cultures, these three cases also show that the change agents for sustainable transformation cannot be short-term visitors and have to spend their entire life working with the community. Dr. Kurien, Dr. Yunus and Father Maria gave undivided time and attention to their organizations. Grameen Bank and Mondragon started with the personal visions of Dr. Yunus and Father Maria, whereas Kurien nurtured AMUL on the vision provided by Tribhuvan Das Patel. In all three cases, it is clear that the change agents were totally committed to the transformation of the community and the organizations emerged as a tool toward achieving that objective. We also find that the change agents were not self-promoting leaders but cause-promoting crusaders—Dr. Kurien championed the "Right to Livelihood," Dr. Yunus championed the "Right to Credit" and Father Maria championed the "Right to Employment." Recent incidents of corruption in multinationals and financial scandals in profit and not-for-profit organizations alike indicate that we have leaders who are more interested in hijacking the organizations for their personal goals rather than steering them toward sustainable transformation of the community. Thus, as supported by the three cases, community transformation seems to require leaders to provide both vision and long-term operational support by being embedded in the community irrespective of cultural differences.

This chapter shows that in the implementation phase of any interventions, it is important that the change agent become an insider rather than a dispassionate external observer. Being an insider does not mean that the change agent will lose his or her objectivity. Instead, by being an insider, the change agent can contribute to the understanding of the issues with the necessary subjectivity, which will help him or her to implement solutions that are culturally appropriate. An insider change agent is similar to an anthropologist studying a culture, the difference being that the change agent uses his or her insights to transform the community with the full involvement of the community. Some of these transformations would require cultural changes, but since the community and the change agent would jointly identify them, it would be both ethical and easy to carry out the implementation.

The five principles demonstrated in this study have four important implications for transformations that are sustainable. First, the three case studies show that learning is bi-directional in that both the agents of change and the communities learn in the process of transforming the community. This social learning, which has been argued as central to non-coercion (Ison, Röling, & Watson, 2007), is important for sustainable development as it makes the change process spontaneous because learning is implicitly motivating and self-propelled and inherently incremental. Social learning also contributes to the learning of the community at the collective level, akin to what has been referred to as organizational learning. Thus, following these five principles creates a learning community, which fires up the transformation process that continues to grow to meet the changing needs of the community.

Second, the three case studies show that sustainable transformations can only be successful when innovative solutions are culturally appropriate, which is often the case when they are derived from the communities. In other words, communities themselves serve as a capital (Callaghan & Colton, 2008) for solutions needed for their transformation. Thus, importing "best practices" must be done cautiously, since they can become a liability if special attention is not given to adapting them to fit the cultural ethos, and may in the long-run hurt rather than help sustainability.

Third, the importance of defining problems with the community and measuring them in multiple ways was highlighted by the three cases. Following the five principles should support sustainable development by creating an environment of trust where continuous communication can take place between change agents and community members. The relevance of doing so has been noted by Holden (2008: 475) in the context of the planning process:

> "[I]f planners and members of the democratic public can work towards a common understanding that it is a process of continuous communication and interaction among citizens and experts that needs to be sustained in the push toward sustainable development, that knowledge is to be generated and tested in public contexts, and that stories have standing alongside scientific models and statistics, important moves toward sustainability can be made in the planning profession as a whole."

Fourth, the five principles not only address the planning process but also the implementation and evaluation phases of any change process. For example, stories serve as qualitative indicators and also provide a thick description that helps in understanding the problems and in finding relevant solutions. Stories also provide identity to the community members and guide future efforts by serving as milestones. Thus, the five principles presented here appear to be inherently supporting sustainable changes in communities.

The model presented in the paper is grounded in theories and ideas from community psychology, and shows what it takes to start a community intervention "of the people, by the people, for the people," to borrow a phrase from Abraham Lincoln's Gettysburg Address given on November 19, 1863. What is interesting is that the three cases presented above amply demonstrate the power of community building when these precepts are practiced. As these cases come from three countries, cultures, and economic sectors, it is a generalizable finding that empowering people does lead to transformation of the community, a community that "shall not perish from the earth," to quote Lincoln again.

This chapter has implications for leadership research and practice. The three social change agents discussed in the paper are transformational leaders of the type that Burns (1978) presented rather than the one that Bass (1990) discussed. These leaders have made personal sacrifices to help their followers, and thus have transformed themselves personally beyond their own imagination as was noted above. The leadership literature is often silent about the transformation in the leaders. We assert that transformation is a two-way street, and this transformation is the reward that keeps the leader or social agent motivated rather than any monetary benefit or gain in social status. This is consistent with Burns' description of how transformational leadership involves both the leaders and the followers transforming each other to higher levels of morality. Burns suggested that the collaborative process and relationship emerges from the appeal to social values, and this is clearly the case with the three social agents discussed in the paper. Thus, the reason for the limited success of the social engineering paradigm is the exchange based interaction between change agents and community members, i.e., the transactional leadership style adopted by change agents, whereas transformational leadership leads a community to constantly reinvent itself.

This study also contributes to the field of organizational development (OD) by suggesting how the OD process can be improved by using the five principles presented above. Organizational development cycle commonly consists of diagnosis, intervention and evaluation.[25] In diagnosing the problem, the process can be improved by not only defining the problem from the perspective of the employees as is commonly done but also by including end users and other stakeholders as well as measuring the problem using multiple methods. Following this approach will help identify and define the problem from multiple perspectives and prevent the neglect of issues considered important by end-users, which is critical for sustainable transformation. In evaluating organizational development efforts, often effectiveness is used as a criterion for the cost sustainability. The principles identified here do not undermine the importance of effectiveness and every effort should be made to increase effectiveness. However, effectiveness should be examined against the criterion of sustainability. It is often in the self-interest of change agents to demonstrate quick results but doing so without emphasizing sustainability may do more harm than good to the community, and no organizational development is worth the effort if it is not going to have a lasting impact on the community. Thus, by adopting these five principles organizational development process can help transform the change process leading to community transformation rather than short-term and tentative changes that breed cynicism and create a negative image of the change process itself.

Finally, it is clear from the three cases that the change agents have to commit their life to achieve such transformations. This is a lesson that community psychologists and social science researchers need to internalize. Researchers cannot expect to have impact by scientifically testing an intervention in the spirit of creating a scientific society and testing reforms as experiments as Campbell (1969) envisioned. What is needed is a lifetime commitment, which would then transform both the community and the researcher. This idea is not new to researchers who know that for research to have a significant impact one needs to work on a program of research rigorously testing various aspects of a theory over many years. Also, researchers generally agree that no research is considered of value if it does not have an impact on the real world. Thus, social scientists need to become humanists who get in the trenches with community members, share their tools and commit to empowering the people in the community to define problems and issues in their own vocabulary and worldview. Community citizens should measure variables of interest in multiple ways, and implement the solutions that they think could help solve the problem. Social scientists need to show their commitment by remaining embedded and passionate facilitators who are ever willing to change the course. There is perhaps not one social science theory that has impacted society and lead to social change to the same degree as any of these three organizations have done. This should assure even the skeptics that it is time for a paradigm shift in social change, and there is a need to move from expert driven interventions to people championed programs that will allow the community to transform and continue to grow spontaneously.

Notes

The authors received the Rupert F. Chilsom Best Theory to Practice Award from the Organizational Development and Change Division of the Academy of Management on August 10, 2009 for this paper.

1. We are grateful to Prof. Jai B. P. Sinha, Dr. Jai Mangal, Serena Wong, and the three anonymous reviewers for their insightful comments that helped us improve the chapter significantly.
2. The term social architecture was introduced by Perlmutter (1965) to develop a theory of social change that could be applied to bring about change in the society, which was grounded in the engineering paradigm of problem solving.
3. Some of the basic information about Amul is available at their webpage (http://www.amul.com/) as well as at http://www.nic.coop/founders/amul.asp.
4. http://www.indiadairy.com/cont_highest_milkproducer_amulorigin.html.
5. http://www.indiadairy.com/cont_highest_milkproducer_amulorigin.html.
6. http://www.indiadairy.com/cont_highest_milkproducer_amulorigin.html.
7 http://www.irma.ac.in/institute/mission.php.
8. AMCUS is used to capture member and product information including milk fat content, the volume collected and amount payable to each member.
9. Food and Agricultural Organization of United Nations: Economic and Social Department: The Statistical Division. http://www.fao.org/es/ess/top/commodity.html?lang=en&item=951&year=2005.
10. http://www.worldfoodprize.org/laureates/Past/1989.htm.
11. http://www.worldfoodprize.org/Laureates/Past/1989.htm.
12. http://www.indiadairy.com/cont_highest_milkproducer_amulorigin.html.
13. http://www.experiencefestival.com/a/Verghese_Kurien_-_The_White_Revolution/id/4996051.
14. Micro Health Plans for Macro Benefit. (2007, January 24). *The Washington Times*, p. A17. Retrieved March 29, 2008, from Questia database.
15. "Unity, work, discipline" (opinion & editorial). *Manila Bulletin*. Manila Bulletin Publishing Corp. 2006. Retrieved June 14, 2009 from HighBeam Research: http://www.highbeam.com/doc/1G1-155756441.html.
16. http://www.grameen-info.org/index.php?option=com_content&task=view&id=22&Itemid=109&limit=1&limitstart=10.

17. http://nobelprize.org/nobel_prizes/peace/laureates/2006/yunus-lecture-en.html.
18. http://www.grameen-info.org/bank/socialbusinessentrepreneurs.htm.
19. Luftwaffe's Top Gun. (2005, December 12). *The Daily Mail* (London, England), Answers to Correspondents. p. 53. Retrieved March 29, 2008, from Questia database.
20. http://www.mcc.es/ing/quienessomos/historia.html.
21. http://www.afsc.org/newengland/bigcat/ttl.php?FID=1208.
22. http://gvanv.com/compass/arch/v1405/saint.html.
23. http://www.mcc.es/ing/contacto/faqs.html#.
24. The loan and its payment is still the responsibility of the individual in the Grameen Model.
25. http://www.managementhelp.org/misc/new_OD.htm.

References

Arisaka, Y. (2003). "Women carrying water: Homeplace, technology and transformation." In Peter D. Hershock, Marietta Stepaniants & Robert T. Ames (eds.), *Technology and Cultural Values: On the Edge of the Third Millennium* (pp. 236-251). Honolulu: University of Hawaii Press.

Auwal, M. A. (1996). "Promoting microcapitalism in the service of the poor: The Grameen model and its cross-cultural adaptation." *The Journal of Business Communication, 33*(1), 27-49.

Banyard, V., & Miller, K. (1998). "The powerful potential of qualitative research for community psychology." *American Journal of Community Psychology, 26*, 485-505.

Barker, C., & Pistrang, N. (2005). "Quality criteria under methodological pluralism: Implications for conducting and evaluating research." *American Journal of Community Psychology, 35*(3/4), 201-212.

Bass, B. M. (1990). "From transactional to transformational leadership: Learning to share the vision." *Organizational Dynamics,* (Winter): 19-31.

Bhattacharyya, J. (2004). "Theorizing Community Development." *Journal of the Community Development Society, 34*(2), 5-35.

Bhawuk, D. P. S. (2008). "Globalization and indigenous cultures: Homogenization or differentiation"? *International Journal of Intercultural Relations, 32* (3), 305-317.

Bhawuk, D. P. S. (2001). "Looking for the silver lining: Integrative potentials in international project negotiation." *Psychology and Developing Societies, 13* (2), 243-262.

Burns, J. M. (1978). *Leadership.* New York: Harper & Row.

Callaghan, E. G., & Colton, J. (2008). "Building sustainable & resilient communities: A balancing of community capital." *Environment, Development and Sustainability, 10*(6): 931-942.

Campbell, D. T. (1969). "Reforms as experiments." *American Psychologist, 24*, 409-429.

Clark, C. M. (2004). "The Mondragon Corporacion Cooperativa: An interview with Juan M. Sinde, chief executive deputy." *Review of Business, 25*(1), 4.

Cooper, C. R., & Denner, J. (1998). "Theories linking culture and psychology: Universal and community-specific processes." *Annual Review of Psychology, 49*, 559-584.

Dale, A., Ling, C., & Newman, L. (2008). "Does place matter? Sustainable community development in three Canadian communities." *Ethics, Place and Environment, 11*(3): 267-281.

Dana, L. P., & France, N. (1996). "Small business in Mozambique after the war." *Journal of Small Business Management, 34*(4), 67.

Dumas, J. E., Rollock, D., Prinz, R. J., Hops, H., & Blechman, E. A. (1999). "Cultural sensitivity: Problems and solutions in applied and preventive intervention." *Applied & Preventive Psychology, 8*, 176-196.

Esman, M.J. (2003). Social Capital and Empowerment. Cornell University.
Esman, M. J. (1972). "The elements of institution building." In J. W. Eaton (ed.) *Institution Building and Development* (pp. 21-39). Beverly Hills, CA: Sage.
Esman, M. J., & Bruhns, F. C. (1966). "Institution building in national development: An approach to induced social change in transitional societies." In Hollis W. Peter (Ed.) *Comparative Theories of Social Change* (pp. 318-342). Ann Arbor, MI: Foundation for Research on Human Behavior.
Ferguson, K. M. (2007). "Implementing a social enterprise intervention with homeless, street-living youths in Los Angeles." *Social Work, 52*(2), 103-112.
Fetterman, D. M., & Wandersman, A. (Eds.) (2004). *Empowerment evaluation principles in practice.* New York: Guilford Press.
Ford, N. (2007, April). "Can the Grameen experience work in Africa? Can Africa learn from the revolutionary micro-finance solutions pioneered by the Grameen bank experience in Bangladesh? " The Bank's Founder and the Bank Itself Were Awarded the 2006 Nobel Peace Prize. Neil Ford Discusses the Philosophy Behind the System. *African Business,* 74.
Foster-Fishman, P.G., Cantillon, D., Pierce, S.J., & Van Egeren, L.A. (2007). "Building an active citizenry: The role of neighborhood problems, readiness, and capacity for change." *American Journal of Community Psychology, 39*(1), 91-106.
Freire, P. (1970). *Pedagogy of the oppressed.* New York: Herder & Herder.
Ganesh, S. R. (1980). "Institution building for social and organizational change: An application." *Organization Studies, 1*(3), 209-228.
Gergen, K. J., Gulerce, A., Lock, A., & Misra, G. (1996). "Psychological science in cultural context." *American Psychologist, 51*, 496-503.
Greenfield, P. M. (1997). "You can't take it with you: Why ability assessments don't cross cultures." *American Psychologist, 52*, 115-1124.
Guarnaccia, P. J., & Rodriguez, O. (1996). "Concepts of culture and their role in the development of culturally competent mental health services." *Hispanic Journal of Behavioral Sciences, 18*, 419-443.
Guillé, M. F. (2001). *The limits of convergence: Globalization and organizational change in Argentina, South Korea, and Spain.* Princeton, NJ: Princeton University Press.
Heller, F., E. Pusic, G. Strauss and B. Wilpert, Eds. (1998). *Organizational participation: Myth and reality.* New York, Oxford University Press.
Herrera, D. (2004). "Mondragon: a for-profit organization that embodies Catholic social thought." *Review of Business,* 25(1), 56.
Herr, K., & Anderson, G. L. (2005). The action research dissertation: A guide for students and faculty. Thousand Oaks, CA: Sage.
Hertz-Bunzl, N. (2006). "Financing hope: Improving microfinance." *Harvard International Review, 27*(4), 32-34.
Holden, M. (2008). "The tough minded and the tender minded: A pragmatic turn for sustainable development planning and policy." *Planning Theory and Practice, 9*(4): 475-496.
Hughes, D.L., & Seidman, E. (2002). "In pursuit of a culturally anchored methodology." In T.A. Revenson, et al. (Eds.). *Ecological research to promote social change: Methodological advances from community psychology* (pp. 243-256). New York: Kluwer Academic/Plenum.
Ison, R., Röling, N., & Watson, D. (2007). "Challenges to science and society in the sustainable management and use of water: Investigating the role of social learning. *Environmental Science and Policy, 10*(6): 499-511.
Julian, D.A. (2006). "A community practice model for community psychologists and some examples of the application of community practice skills from the partnerships for success initiative in Ohio." *American Journal of Community Psychology, 37*(1), 21-28.

Kanter, R.M. (1972). *Commitment and community: Communists and utopias in sociological perspective.* Cambridge, Mass.: Harvard University Press.

Kaplan, A. (2002). *Development practitioners and social process—Artists of the invisible.* London: Pluto Press.

Kelly, J.G., Ryan, A.M., Altman, B.E., & Stelzner, S.P. (2000). "Understanding and changing social systems: An ecological view." In J. Rappaport & E. Seidman (Eds.), *Handbook of community psychology* (133-160). New York: Kluwer, Academic, Plenum.

Kelly, J.G. (2007). "The system concept and systemic change: Implications for community psychology." *American Journal of Community Psychology, 39*(4), 415-419.

Kurien, V. (2001). The Amul Saga. Seminar: Markets in Motion. Vol. 498. http://www.india-seminar.com/2001/498/498%20verghese%20kurien.htm.

Langhout, R.D. (2006). "Where am I? Locating myself and its implications for collaborative research." *American Journal of Community Psychology, 37*(4), 267-274.

Lincoln, A. (1863). Draft of the Gettysburg Address, Nicolay Copy.

Lutz, M. A. (1993). "Cardinal issues in the future of social economics: A humanistic view." *Review of Social Economy, 51*(4), 455-75.

Marsella, A. J. (1998). "Toward a "global-community" psychology: Meeting the needs of a changing world." *American Psychologist, 53*, 1-10.

Miller, A. (1996). "Values for new economic relationships." *The Ecumenical Review, 48*(3), 317.

Mohatt, G.V., Hazel, K.L., Allen, J., Stachelrodt, M., Hensel, C. & Fath, R. (2004). "Unheard Alaska: Culturally Anchored Participatory Action Research on Sobriety with Alaskan Natives." *American Journal of Community Psychology, 33*(4), 263-274.

Mondragon Corporation Cooperative (July, 2006). *The history of an experience.* Retrieved March 29, 2008, from http://www.mcc.es/ing/quienessomos/historia.html.

Morrison, R. (1991). *We build the road as we travel.* Philadelphia, PA: New Society Publishers.

Pena, D., & Gallegos, J. (1997). "Local knowledge and collaboration: Environmental action research." In P. Nyden, A. Figert, M. Shibley, & D. Burrows (Eds.), *Building community: Social science in action* (pp. 85-91). Thousand Oaks, CA: Sage.

Perlmutter, H. (1965). *Towards a theory and practice of social architecture.* London: Tavistock.

Peters, T. J., & Waterman, R. H. (1982). *In search of excellence: Lessons from America's best-run companies.* Cambridge, MA: Harper & Row Publishers.

Prahalad, C.K. and Hart, S.L (1998). "The Fortune at the Bottom of the Pyramid." *Strategy+Business, 26*: 54-67.

Price, R.H. & Behrens, T. (2003). "Working Pasteur's quadrant: Harnessing science and action for community change." *American Journal of Community Psychology, 31*(3/4), 219-223.

Rappaport, J. (1981). "In praise of paradox: A social policy of empowerment over prevention." *American Journal of Community Psychology, 9*, 1-26.

Roger, P. (2007). *Dr. Kurien speaks: An interview* (English Version: 35 minutes).Interview Accessed from: http://www.amul.tv/amultv/viewfeaturedmovie_kurienenglish.php?version=English.

Russell, R. A., & Rus, V. A. (Eds.). (1991). *International handbook of participation in organizations for the study of organizational democracy, co-operation, and self-management (vol. 2).* Oxford: Oxford University Press.

Sarason, S.B. (2003). The Obligations of a Moral-Scientific Stance. *American Journal of Community Psychology,* 31, 209-211.

Schram, B. (1997). *Creating small scale social programs: Planning, implementation, and evaluation.* Thousand Oaks, CA: Sage.

Schorr, L. B. (1997). *Common purpose: Strengthening families and neighborhoods to rebuild America.* New York: Anchor Books.

Sen, A. (2005). "How does development happen?" *The Cato Journal, 25*(3), 455-459.

Sherif, M., & Sherif, C. (1959). *An outline of social psychology.* New York: Harper and Row.

Shipp, S. C. (1996). "The road not taken: Alternative strategies for black economic development in the United States. *Journal of Economic Issues, 30*(1), 79-95.

Shweder, R. (1995). "Cultural psychology: What is it?" In N. R. Goldberger & J. B. Veroff (Eds.) *The culture and psychology reader* (pp. 41-86). New York: New York University Press.

Siffin, W.J. (1972). "Institution building as vision and venture: A critique." In J.W. Eaton (ed.) *Institution building and development* (pp. 41-64). Sage: Beverly Hills.

Snow, D.L. (2000). "The development of a community psychology setting: Integration of service, research, and training." In J. Rappaport & E. Seidman (Eds.), *Handbook of community psychology* (pp. 748-752). New York: Kluwer, Academic, Plenum.

Speer, P.W., & Hughey, J. (1995). "Community organizing: An ecological route to empowerment and power." *American Journal of Community Psychology, 23*, 729-748.

Spoth, R.L., & Greenberg, M.T. (2005). Toward a Comprehensive Strategy for Effective Practitioner-Scientist Partnerships and Larger-Scale Community Benefits. *American Journal of Community Psychology, 35* (3/4), 107-126.

Stein, C.H., & Mankowski, E.S. (2004). "Asking, witnessing, interpreting, and knowing: Conducting qualitative research in community psychology." *American Journal of Community Psychology, 33*(1), 21-36.

Stewart, E. (2000). "Thinking through others: Qualitative research and community psychology." In J. Rappaport & E. Seidman (Eds.), *Handbook of community psychology* (725-736). New York: Kluwer, Academic, Plenum.

Stoecker, R. (1997). "The imperfect practice of collaboration research. The 'Working Group on Neighborhoods' in Toledo, Ohio." In P. Nyden, A. Figert, M. Shibley, & D. Burrows (Eds.), *Building Community: Social Science in Action* (pp. 219-225). Thousand Oaks, CA: Sage.

Stryjan, Y. (1989). *Impossible organizations: Self-management and organizational reproduction.* New York: Greenwood Press.

Taylor, F. W. (1911). *The principles of scientific management.* New York, NY: Harper and Brother Publishers.

Tharp, R. G., & Gallimore, R. (1982). "Inquiry process in program development." *Journal of Community Psychology, 10*, 103-118.

Thomas, H. & Logan, C. (1982). *Mondragon: An economic analysis.* Boston, MA: G. Allen & Unwin.

Triandis, H. C., & Bhawuk, D. P. S. (1997). "Culture theory and the meaning of relatedness." In P. C. Earley & M. Erez (Eds.), *New perspectives on international industrial/organizational psychology,* (pp. 13-52). New York, NY: The New Lexington Free Press.

Trickett, E. J. (1984). "Towards a distinctive community psychology: An ecological metaphor for training and the conduct of research." *American Journal of Community Psychology, 12*, 261-279.

Trickett, E.J., & Espino, S.L.R. (2004). "Collaboration and social inquiry: Multiple meanings of a construct and its role. *American Journal of Community Psychology, 34*(1), 1-76.

van Uchelen, C. (2000). "Individualism, collectivism, and community psychology." In J. Rappaport & E. Seidman (Eds.), *Handbook of community psychology,* (65-78). New York: Kluwer, Academic, Plenum.

Vincec, S. & MacLeod, G. (1996). "Don José Maria Arrizmendi-Arrieta: Mondragon pastor made ideas real." *Compass: A Jesuit Journal, 14*(5), 23.

Wandersman, A. (2003). "Community science: Bridging the gap between science and practice with community-centered models." *American Journal of Community Psychology, 31*(3/4), 227-242.

Wandersman, A., & Florin, P. (2000). "Citizen participation and community organizations." In J. Rappaport & E. Seidman (Eds.), *Handbook of community psychology* (247-272). New York: Kluwer, Academic, Plenum.

Weisner, T. S. (1996). "Why ethnography should be the most important method in the study of human development." In R. Jessor, A. Colby, & R. A. Shweder (Eds.), *Ethnography and human development: Context and meaning in social inquiry,* (pp. 305-326.). Chicago: University of Chicago Press.

Whyte, W. F. & Whyte, K. K. (1991). *Making Mondragon: The growth and dynamics of the worker cooperative complex.* (2nd ed.). Ithaca, NY: ILR Press.

Wicker, A. W. (1987). "Behavior settings reconsidered: Temporal stages, resources, internal dynamics, and context." In D. Stokols & I. Altman (Eds.), *Handbook of environmental psychology* (pp. 613-653). New York: Wiley.

Yunus, M. (1997, July/August). "Empowerment of the poor: Eliminating the apartheid practiced by financial institutions." *The Humanist, 57*, 25. Retrieved from FindArticles.com. 12 Sep, 2009. http://findarticles.com/p/articles/mi_m1374/is_n4_v57/ai_19582386/.

Zimmerman, M.A. (2000). "Empowerment theory: Psychological, organizational, and community levels of analysis." In J. Rappaport, & E. Seidman (Eds.). *Handbook of community psychology.* New York: Kluwer, Academic, Plenum.

Sustainable Globalization and Limits to Sustainability[1]

Imre Lévai

Introduction

Immanuel Wallerstein, the American historian-sociologist elaborated his world system theory originally without any direct or indirect reference to the general or complex systems theories (Wallerstein, 1974).[2] For quite a long time he remained within the framework of the Aristotelian view of whole-part relationships, while—at the same time—strove to apply the method of the Hegelian dialectics, having advanced the consequences of complex systems behavior some two centuries ago. It was perhaps his inadequate knowledge of political economy for, while he was thoroughly scrutinizing the general issues of social division of labor (enchanting sociologist at large), he did not pay particular attention to the structural and functional aspects of accumulation of capital. *Wallerstein* left no doubt, however, that he saw the *modern* (complex) world system emerging, and the beginning of globalization in the sense of non-linear dynamical system evolution, in the second half of the fifteenth century, the period associated with "antediluvian" commercial (more precisely: merchant's) capital penetrating the productive spheres of the economy (Wallerstein, 1974, 2000).

Though till the mid-1700s this evolutionary process continued at a slow pace, the positive feedback effects of industrial revolution on society involved the transition from formal to real subsumption of labor under capital, from original (primitive) to organic (specifically capitalist) accumulation of capital (Marx, 1990)—with the unequal exchange consequences of spatial and temporal asymmetries in the world system.[3]

Complexity of Globalization

On January 23, 2000 Stephen Hawking stated in San Jose Mercury News:

"I think the next century will be the century of complexity."

Several authors, over-generalizing the Wallersteinian paradigm, visualize the emergence of the world system not 500 but 5000 years ago (Frank–Gills, 1993), thus interpreting the modern phenomenon of globalization in a particular historicist way. No doubt, globalization or the evolution of the world system (Modelski, 2000; Lévai, 2006) are processes determining the major course of the historical evolution of humankind and in this sense they can be traced back to the swarming out of the human race from Africa, and from then on, all empires and all religions, surpassing the frontiers of one continent, can be seen as early manifestations of these processes and phenomena. In pre-capitalist or pre-modern ages the Holocene environmental system constituted the genuine global system for humankind; the interrelations between local societies annexed by world empires or converted to common faith, however, remained contingent and unsystematic. If, however, the process of globalization is viewed as the evolution of a system of social relations extending over the globe, as the emergence of a complex worldwide social system, then the specific system-creating and integrating social force and its function should be defined. In the case of the modern world system that force is capital and its function is accumulation.

The integrating and differentiating effects (convergent and divergent tendencies, Lévai, 1984) of globalization are such complex social phenomena, involving multiple mutual determinations and positive and negative feed-back/feed-forward effects among the sub-systems of the world system, the current understanding and proper interpretation of which would require the application of the most modern and appropriate means of scientific inquiry from students of humanities in the broad sense. To view the inseparable aspects of socio-economic structure and function as a whole, conditioned and determined by division of labor and accumulation of capital, respectively, was made possible by the advent and intervention of the (not so) new science of complexity.

According to the sadly neglected report by the *Gulbenkian* Commission, set up by outstanding scholars of various disciplines from all over the world:[4]

> The importance of complex systems analysis for the analysis of social science is far-reaching. Historical social systems are quite clearly composed of multiple, interacting units, characterized by the emergence and evolution for nested hierarchical

organization and structure, and complex spatiotemporal behavior. Furthermore, in addition to the kind of complexity exhibited by nonlinear dynamic systems with fixed, microscopic mechanisms of interaction, historical social systems are composed of individual elements capable of internal adaptation and learning as a result of their experience. This adds a new level of complexity (one which is shared with evolutionary biology and ecology) beyond that of the nonlinear dynamics of traditional physical systems. [...] The conceptual framework offered by evolutionary complex systems as developed by the natural sciences presents to the social sciences a coherent set of ideas that matches long-standing views in the social sciences, particularly among those who have been resistant to the forms of nomothetic analysis inspired by the science of linear equilibria. Scientific analysis based on the dynamics of nonequilibria, with its emphasis on multiple futures, bifurcation and choice, historical dependence, and, for some, intrinsic and inherent uncertainty resonates well with important traditions of the social sciences (Juma *et al.*, 1996: 63–64).

Wallerstein and other thinkers in the line of world system and world complexity analysis—e.g., *Samir Amin* (1992) and *James Rosenau* (1990, 1997)—eventually shrank back from deploying the full armory of complexity theory and contented themselves with applying the standard notions of bifurcation, chaos, and turbulence as metaphors to apprehend the complex phenomena of the evolution of the world system.[5] *John Urry*, making a pioneering attempt at grasping the phenomenon of global complexity recently (Urry, 2003), resorts to the implied comparisons of networks and fluids. Others, however, such as *Rosenau* referred to above, turning a blind eye to the complex (non-linear dynamical) differentiating and integrating effects of globalization, fabricated the concept of "fragmegration" from the words fragmentation and integration. The essentially dialectical relationship that the universal manifests in the particular and the global in the local—and *vice versa*—was aptly and perhaps most wittily formulated by the new business term of "glocalisation" picked up from *The Oxford Dictionary of New Words* and used happily by *Roland Robertson* (1995) and many others. By the same token, *Wallerstein* (1984: 167), discussing the contradictory nature of nationalism of the modern world and capitalist civilization in a study written in 1978, also spoke about "universalism through particularism and particularism through universalism."

Coming to the end of this section, it will be apt to quote the Chairman of the *Gulbenkian* Commission, a somewhat exaggerating but most inspiring—and obviously inspired by *Ilya Prigogine* and *Isabelle Stengers* (1979, 1984)—*Wallerstein* on the potential interpretative power of what he calls "new science," that is complex systems analysis:

The deep strength of this rebellion against the centrality of linear, equilibrium processes in scientific analysis and against the theoretical possibility of total precision lies not in any rejection of the basic scientific enterprise—the optimal comprehension

> of material reality—but in the rapprochement of the scientific method (reinterpreted as the attempt to interpret complexity rather than the attempt to reduce complexity to minimalism) with intelligent work in the social sciences and the humanities. It is precisely the denial of the concept of 'two cultures'—science and/versus the humanities—which is the essential thrust. The concept of the two cultures served the same purpose (indeed was essentially the same thing) as the distinction between nomothetic and idiographic epistemologies in the social sciences ... The 'new science' fatally eroded the premises of Baconian-Newtonian science ... however little aware of this some of the 'new scientists' may themselves be (Wallerstein, 1991a: 13).

Later, contemplating on the historical (path-dependent) evolution, and referring to the inherent non-linear dynamics (complexity), of the modern world system, he wrote:

> [T]he capitalist world-economy constitutes an historical system, which thereby has an historical life: it had a genesis; it has a set of cyclical rhythms and secular trends that characterize it; it has internal contradictions which will lead to its eventual demise. [T]he short-run contradictions lead to middle-run solutions which translate into long-run linear curves approaching asymptotes. As they approach these asymptotes, the pressures to return to equilibria diminish, leading to ever-greater oscillations and a bifurcation. Instead of large random fluctuations resulting in small changes in the curve, small fluctuations will result in large changes. The imminence of bifurcation, brought about by the fact that middle-run solutions to short-run contradictions are no longer readily available, is disastrous for the system. [...] This fact is not *necessarily* good news for the opponents of the system. Bifurcations are unpredictable in their outcomes. What will replace the capitalist world-economy eventually may be better, but it could be worse. Thus we have reason neither for despair nor for celebration. Nonetheless, there is one fundamental encouraging element. The outcome will be the result of our collective effort, largely as expressed in the work of the new antisystemic movements. Whereas before large fluctuations resulted in small changes (hence the 'determinism,' hence the disillusionment with the outcomes of reformism, even when called 'revolution'), now small fluctuations will result in big changes (hence the opening for true 'agency,' hence the responsibility we all bear.) The world-system is in mutation now. This is no longer a moment of the minor, constant cumulation of cycles and trends. 1989 is probably a door closed on the past. We have perhaps arrived now in the true realm of uncertainty. The world-system will continue, of course, to function, even function 'well.' It is precisely because it will continue to function as it has been functioning for 500 years, in search of the ceaseless accumulation of capital, that it will soon no longer be able to function in this manner. Historical capitalism, like all historical systems, will perish from its successes not from its failures (Wallerstein, 1991a: 14-15).

Sustainability of Globalization

"[A]fter so many failed prophecies, is it not in the interest of social science to embrace complexity, be it at some sacrifice of its claim to predictive power"?
—(Hirschman 1992: 139)

No doubt, the future of the modern world system is open and contingent (unpredictable). Because of the high degree of its immanent complexity, only the borderline scenarios at the constraining limits of its possible state space can be projected up to a point of bifurcation (Lévai, 1996, 1998). Among the several attempts at such projections, the first report to the Club of Rome (Meadows et al., 1972) appears the best known and, at the same time, the most controversial. The authors of the report, published just a year before the memorable first oil price shock in 1973, as if having foreseen the descending curve of the world economic cycle and the raw materials and energy crises of the 1970s, warned of the ecological limits to exponential economic growth. The model, applied by *Donella* and *Daniel Meadows* and their colleagues, confronted extrapolations of growth trends of a homogeneous world economy with the finite natural resources of the earth, exploitable at the achieved level of technological development. Upon noting the obvious ecological limits to meeting the soaring demand generated by unlimited exponential growth, the creators of the model put forward proposals to restrict global rates of economic growth.

Apart from some interpreters' and critiques' misinterpretations and misunderstandings, the report soon became the subject of heavy and due criticism. Regardless of the minor weaknesses of the model, the major fault of the "reporters" was their blindness to the complexity of the world economic system. Undoubtedly, all formal models cannot but simulate the real world in a simplified manner, but one should not lose sight of the non-linear dynamics of global evolution when trying to detect future trends in the world system. You may repeat the mantra of "complex nature of global problems" as many times as you wish, but if you do not consider the profound consequences of it when inquiring into the origin and nature of that complexity, you will be misled by the inevitable simplifications (such as the presumption of a homogeneous world economy, of the given level of technological development and mere extrapolations of prevailing trends). Non-linear and dynamical systems are, unfortunately, difficult to handle, but not completely unperceivable and unmanageable.

Some twenty years after the first report on the limits to growth had come to light *Donella Meadows* still did not seem to arrive at terms with her critiques and, worse enough, with complexity:

> There are limits—environmental, economic, social. There is a limit to the amount of oil under the ground and a limit to the amount of carbon dioxide the atmosphere can hold before the planet warms, the seas rise, and the climate changes. There is a limit to the amount of injustice people will tolerate. There is a limit to the ability of human beings to manage complex systems (Meadows, 1991: 30).

It is all the more surprising as the sixth report to the Club of Rome on "no limits to learning" made a crucial contribution to the clarification of the complex nature of what the authors called the "world problematique" as early as in 1979. Since the three reporting scholars, having represented the "three worlds" of that era, sent a valuable message on the history of human cognition,[6] it is worth quoting at some length:

> It is possible to read the history of humanity as a sustained effort to overcome complexity through increasingly refined and effective means first of representing reality and then of acting upon it. Developments in knowledge, technology, power, organization, norms of conduct, and above all the creation of coherent mental constructs to represent the surrounding environment have resulted from the interplay between the challenge of complexity and the urge to master it. This process is nothing new. Whenever a problem has fallen outside the boundaries of some tested procedure to deal with it, the situation was labelled 'complex.' Today, the oldest sources of complexity, namely the universe and Nature, still continue to pose a bewildering number and variety of facts and factors which astronomers, biologists, and other scientists try to render intelligible by new theories and concepts such as black holes or the deciphering of the genetic codes ... (Botkin *et al.,* 1979: 17-18).

The authors go on disclosing the complex issue of accelerating global social complexity ("second order complexity") inherent in what they call the world problematique:

> There is another type of complexity of more immediate interest than that engendered by natural systems—a second order complexity caused by human actions and man-made systems, and represented by a world of culture, civilization, and human artefacts. Contemporary societies face a sudden, menacing, intensification of this more recent complexity in virtually all fields. An accelerating tempo of change, a sheer increase in numbers and in size, deepening uncertainties, and extreme risks such as those inherent in the world problematique are integral aspects of the new complexity which signify a change in kind as well as in degree. There are many sources of this complexity. One is our limited ability, despite the assistance of computers, to deal effectively and simultaneously with large number of factors. Other sources of complexity go beyond an increase in numbers, such as the imprecision of a world which is not amenable to a simple 'yes and no' logic; the ambiguity of language which leaves agreed-upon international documents open to varying interpretations; the uncertainty of future events which renders obsolete the old mental habit of ignoring long-term implications; and finally the search for the globality and the whole. This last factor is to be understood not only in the widening horizons of concern implied by 'interdependence,' but in the need to grasp new interconnections among phenomena and people and to develop a better capacity for synthesis. In the field of ecology, for example, we were caught methodologically unprepared to deal with the counter-intuitive behaviour of large systems whose deceptive resilience, once exceeded, can give way to discontinuity, rupture, and even catastrophe. Two basic ways to reduce complexity can be envisaged. The first is to attempt to simplify reality. Much of the work of science is aimed at formulating simple hypotheses and creating powerful concepts to increase our understanding. This approach all too often entails the pitfall of slipping into 'reductionism' where simplifying concepts are reduced to over-simplifications. The second strategy is to 'absorb' complexity by differentiating,

restructuring, and improving our means to cope with it. It is this process we have in mind when we speak of learning. Through learning, individuals and societies can develop the capacity to face new situations of growing complexity. This approach harmonizes humanity with a pervasive trend in nature to progress from simple units to ever more complex configurations. With human action and societies more complex today than in the past, learning must speed up its pace to avoid lagging behind the objective processes or independent events of nature and society. It is this current shortfall, which is what we termed the human gap, that makes complexity seem overwhelming and unmanageable (Botkin *et al.*, 1979: 18-19).

The sixth report was not silent concerning the ecological hazards threatening the evolution of the global human society either, but it raised the issue in a context quite different from that of the first one:

> The major problem is not, as generally thought, the depletion of non-renewable resources. Although over-exploited, these can still be found in respectable quantities in the Earth's crust and oceans. Nevertheless, some resources are becoming either physically scarce, or more expensive to extract and process, or both. Farmland and energy, for example, are in these borderline categories. Once wars were waged in quest for salt; today energy is, and tomorrow food may become, the salt of the contemporary economy. However, the so-called renewable resources face more imminent dangers; among these are rapid degradation of the world's tropical rain forests, the advance of desertification, and an accelerating extinction of animal and plant wildlife. If these trends continue, we are destined to lose drastically in terms of habitat, health, and quality of life, if not even the very capacity for survival. This is the chain of cold facts and actual trends we are witnessing at this fateful crossroad to the future. They are of such a magnitude and nature, and their interactions so critical, that everything human is upset and made immensely more hazardous by their complexity. For nearly ten years now, we have recognized the dangers posed by the world problematique; yet the human condition continues to deteriorate and our understanding continues to lag, despite all our scientific knowledge, all our educational achievements, and all our research capabilities (Botkin *et al.*, 1979: 3-4).

The authors of the first report, however, do not seem to be able to get rid of their Malthusian (Malthus, 1970) obsession with extrapolating exponential rates of growth. Now thirty years after having published their "Limits to Growth", in an "Update" they refer to the requirement of the sustainability of social evolution and call attention to the need of mechanisms of controlling positive feedback (self-reinforcing) effects:

> From a systems point of view, a sustainable society is one that has in place informational, social, and institutional mechanisms to keep in check the positive feedback loops that cause exponential population and capital growth (Meadows *et al.*, 2002: 254).

With a view to sustainable global evolution, the *Meadowses* and *Randers* (*Behrens* not involved this time) emphasize that, contrary to the various misinterpretations of their "limits" thesis of 1972, they did not visualize a "zero rate of growth" strategy, that is the alternative of worldwide stagnation, then and they do not do so now either. The rationally and feasibly sustainable path of evolution, in their opinion, should be found by heading towards the stationary state, on a global scale, proposed by *John Stuart Mill* more than one and a half decades ago.[7] Let us disregard the not too attractive alternative of globally extending a pattern of social involution (a kind of "Asiatic mode of production") that would deprive the whole social world system itself from the capability for adapting to the changes in the natural environment by reducing the complexity (the general evolutionary and the specific emergent attributes) of its differentiated structural and functional sub-systems, and let us pay attention to the optimistic words of the messengers of a brave new world:

> The word *equilibrium* in systems language means that positive and negative loops are in balance and that the system's major stocks—in this case population, capital, land, land fertility, nonrenewable resources, and pollution—are held fairly steady. It does *not* necessarily mean that the population and economy are static or stagnant. They stay roughly constant in total size, the way a river stays roughly constant in volume, even though water is always running through it. In an 'equilibrium society' ... some people are being born while other are dying; new factories, roads, buildings, machines are being built while old ones are being decommissioned and recycled. While technologies are improving, the flow of material output per person would almost certainly be changing in form, diversifying in content, increasing in quality. As a river may rise and fall around some average flow, so could an equilibrium society vary, either by deliberate choice or by unforeseen opportunities or disasters. As a river can purify itself and support more rich and varied aquatic communities when its pollution load is diminished, so a society can purify itself of pollution, acquire new knowledge, make its production processes more efficient, shift technologies, improve its own management, make distribution more equitable, learn, and evolve. We think society is more likely to do all those things when the strains of growth are alleviated and when it is changing slowly enough that there is time for full understanding, reflection, and choice about the effects of its decisions (Meadows *et al.*, 2001: 246-247).

The above image of, and arguments for, an idyllic stationary and sustainable "equilibrium" state are built on serious misconceptions of systems theory. It must be clearly seen that positive and negative loops will never be in "balance" in an open non-linear dynamical system such as human society, in general, and world society, in particular. This is also why a complex system's "major stocks" cannot be held steady (Eve, 1997: 280). The only possible alternative is a self-sustained path of world system evolution within a (steady) state space (zone of attraction), limited by the initial (historical) conditions of its functioning, from which even a relatively minor (disproportionate to the effect) change in the external (environmental) or in the internal (sub-systemic) conditions may divert it. An open complex (non-linear dynamical) system cannot and will not *per definitionem* be in equilibrium since it is always at the edge of chaos (Eve *et al.,* 1997). To put it figuratively: the water is not running through the river, since the running water is the river and its streams and whirlpools are shaping its structure right from its spring, and its structure is the system of its irreversible streams and path-dependent whirlpools. (To make the image somewhat awkward but complete: the riverbank and bed are the border-line environment.) Any intervention in the functioning of a complex system, however benevolent it appears to be, may result in unexpected effects, not to mention counter-effects, and the all-too-well-known side effects in the fields of e.g., medicine and environmental protection. Changing ("purifying") the state of a society (of a polluted river) is such an adaptive and self-organizing process that would structurally and functionally alter the interdependent and interacting system parameters.

This is what the frequently cited authors of "No Limits," *James Botkin, Mahdi Elmandjra* and *Mircea Malitza,* stressed when they tried to open the eyes of humankind to the complexities it created and, at the same time, broke new paths to bridge the human gap:

> Humanity is entering a period of extreme alternatives. At the same time that an era of scientific and technological advancement has brought us unparalleled knowledge and power, we are witnessing the sudden emergence of a 'world problematique'—an enormous tangle of problems in sectors such as energy, population and food which confront us with unexpected complexity. Unprecedented human fulfilment and ultimate catastrophe are both possible. What will actually happen, however, depends on another major—and decisive—factor: human understanding and action. [...] Our reluctance to face up to unpleasant realities blurs the fact that the current general crisis will get far worse before it can eventually get better. The few existing evaluations or forecasts are narrow, fragmented, or short-term. Never is our vast assortment of resources mobilized across disciplines and national boundaries with a view to pursuing common, global goals on a long-term basis. As a consequence, humanity is pitifully unprepared to cope with the formidable challenges, threats, and complexities on the immediate horizon. We must come to understand at least two critical points. One is that humanity as a whole is moving rapidly towards a momentous crossroads where there will be no room for mistakes. The second is that we must break this vicious circle of increasing complexity and lagging understanding while it is still possible to exert influence and some control over our own destiny and future (Botkin *et al.,* 1979: 1-2).

Some Concluding Remarks: Political Economy of Complexity and Sustainability

In 1979 Ilya Prigogine and Isabelle Stengers wrote on the "Meta-Morphosis of Science":

> *"Notre problème n'est donc pas de réduire le vivant à une description unique, mais d'apprendre à décrire l' 'économie politique' des processus naturels, d'apprendre*

comment l'énergie, la matière, les informations sont emmagasinées, transformées, distribuées. Et nous avons tout lieu de penser que l'économie politique de la nature sera très éloignée des calmes modèles de division du travail et de gestion harmonieuse et centralisée auxquels l'idée d'organisme a été longtemps associée."

"Our problem is not so reduce life to a unique description, but to learn how to describe the natural processes of the 'political economy,' to learn how energy, matter, and information, are stored, processed, and distributed. And we have every reason to believe that the political economy of nature will be very far from a calm model, working harmoniously through a centralized management division, with which the idea of the body was long associated."

Ever since the sixth report to the Club of Rome became public, several attempts have been made to cope with the growing complexity of relationships between the global social and natural systems. Suffice it to recall the dynamical approach of the Report by the World Commission on Environment and Development to the UN General Assembly (*Our Common Future,* 1987, Oxford: Oxford University Press) on sustainable development that "meets the needs of the present without compromising the ability of future generations to meet their own needs." Genuine historical comparative analyses on a global scale are, however, badly needed (Chew, 2006). Whereas the complex properties of natural systems (climate, tectonics, flora and fauna, etc.) and of social systems are widely and deeply discussed separately, the non-linear interactions between the global natural environment and the world social system are still poorly understood (Prew, 2003) and if serious inquiries into the complexities and consequences of such interactions are done, they—because of their sophisticated language they use—remain within the limited circles of academic readers (e.g., Luhmann, 1989). Moral—however honest and moving—appeals to mankind, urging new goals and planetary consciousness (Laszlo, 1977, 1997), would hardly transform individual and institutional patterns of behavior systematically—the global system is to be transformed politically and economically in order to achieve sustainability by comprehending and maintaining social complexity:

From the standpoint of a higher economic form of society, private ownership of the globe by single individuals will appear quite as absurd as private ownership of one man by another. Even a whole society, a nation, or even all simultaneously existing societies taken together, are not the owners of the globe. They are only its possessors, its usufructuaries, and, like *boni patres familias*, they must hand it down to succeeding generations in an improved condition (Marx, 1977 [1894]: 776).

Notes

1. Revised and enlarged version of my paper presented at the international conference on "Globalisation and Sustainable Development" organised by the King Sigismund College and the Executive Committee of EADI in Budapest on 3 April, 2008. I feel obliged for the useful comments by the participants on the original version of it. However, the usual caveats apply.
2. The author would surely protest against the term of "theory" as he insists on the more general notions of "perspective" or "analysis"; cf. Wallerstein 1991b: 237-256.
3. For my earlier attempts at grasping the problem see Lévai 1978, 1986, 1988, 1989, for my recent contributions see Lévai 2006 and 2007.
4. *Immanuel Wallerstein* (USA), chair of the Commission, sociology; *Calestous Juma* (Kenya), science and technology studies; *Evelyn Fox Keller* (USA), physics; *Jürgen Kocka* (Germany), history; *Dominique Lecourt* (France), philosophy; *Valentin Y. Mudimbe* (Zaire), Romance languages; *Kinhide Mushakoji* (Japan), political science; *Ilya Prigogine* (Belgium), chemistry; *Peter J. Taylor* (UK), geography; *Michel-Rolph Trouillot* (Haiti), anthropology.
5. An encouraging exception to the rule was Straussfogel in 1997.
6. The American, Rumanian and Moroccan authors specified the peculiarity of the gap between the growing objective complexity of the world and the subjective capability of man to manage it in the following way:

   "The human gap is the distance between growing complexity and our capacity to cope with it. Clearly, one eternal human endeavour has been to develop additions to knowledge and improvements in action to deal with a complexity which, for most of history, derived primarily from natural phenomena. An essential difference today is that contemporary complexity is caused predominantly by human activities. We call it a *human gap*, because it is a dichotomy between a growing complexity of our own making and a lagging development of our own capacities. Global problems, currently the chief manifestations of complexity, are first and foremost human problems. They are only secondarily attributable to natural causes. As human problems, they inherently encompass all our frailties and potentials. We are not certain whether the issues we identify are complete, correct, or correctly stated. We are still unable to properly assess and respond to the dangerously high levels of risk intrinsic to the world problematique. And it is not only our capacity to cope which is in question but also our ability or willingness to perceive, understand, and take action on present issues as well as to foresee, avert, and take responsibility for future ones"(Botkin et al. 1979: 6-7).
7. "It must always have been seen, more or less distinctly, by political economists, that the increase of wealth is not boundless; that at the end of what they term the progressive state lies the stationary state, that all progress in wealth is but a postponement of this, and that each step in advance is an approach to it. We have now been led to recognize that this ultimate goal is at all times near enough to be fully in view; that we are always on the verge of it, and that if we have not reached it long ago, it is because the goal itself flies before us. The richest and most prosperous countries would very soon attain the stationary state, if no further improvements were made in the productive arts, and if there were a suspension of the overflow of capital from those countries into the uncultivated or ill-cultivated regions of the earth. The impossibility of ultimately avoiding the stationary state –his irresistible necessity that the stream of human industry should finally spread itself out into an apparently stagnant sea—must have been, to the political economist of the last two generations, an unpleasing and discouraging prospect;

for the tone and tendency of their speculations goes completely to identify all that is economically desirable with the progressive state, and with that alone. [...] I cannot, therefore, regard the stationary state of capital and wealth with the unaffected aversion so generally manifested towards it by political economists of the old school. I am inclined to believe that it would be, on the whole, a very considerable improvement on our present condition. I confess I am not charmed with the ideal of life held out by those who think that the normal state of human beings is that of struggling to get on; that the trampling, crushing, elbowing, and treading on each other's heels, which form the existing type of social life, are the most desirable lot of human kind, or anything but the disagreeable symptoms of one of the phases of industrial progress" (Mill 1987 [1848]: 746, 748).

References

A. Gergely, A.—Lévai, I. (eds.) (2004): *Regions and Small States in Europe: Limits to EU Enlargement. Integration Studies No. 16* in the Memory of Éva Táll. Budapest: Research Centre of Ethno-regional Studies at the Institute for Political Science of the Hungarian Academy of Sciences.

Amin, S. (1992): *Empire of Chaos*. New York: Monthly Review Press.

Botkin, J.W.—Elmandjra, M.—Malitza, M. (1979): *No Limits to Learning. Bridging the Human Gap. A Report to the Club of Rome.* Oxford—New York—Toronto—Sydney—Paris—Frankfurt: Pergamon Press.

Chew, S.C. (2006): Dark Ages. Ecological Crisis Phases and System Transition. In: Gills—Thompson, 2006: 163-202.

Denemark, R.A.—Friedman, J.—Gills, B.K.—Modelski, G. (eds.) (2000): *World System History. The Social Science of Long-Term Change.* London—New York: Routledge.

Dunaway, W. A. (2003): *Emerging Issues in the 21st Century World-System. Vol. II. New Theoretical Directions for the 21st Century World-System.* Westport, CT—London: Praeger Publishers.

Eve, R.A.—Horsfall, S.—Lee, M.E. (eds.) (1997): *Chaos, Complexity, and Sociology. Myths, Models and Theories*. London—New Delhi—Thousand Oaks: SAGE Publications.

Eve, R.A. (1997): Afterword: So Where Are We Now? A Final Word. In: Eve *et al.* 1997: 269-280.

Featherstone, M.—Lash, S.—Robertson, R. (eds.) (1995): *Global Modernities.* London—New Delhi—Thousand Oaks: SAGE Publications.

Frank, A.G.—Gills, B.K. (eds.) (1993), *The World System. Five Hundred Years or Five Thousand?* London—New York: Routledge.

Gills, B.K.—Thompson, W.R. (eds.) (2006): *Globalization and Global History.* Oxon—New York: Routledge.

Hirschman, A.O. (1992): *Rival Views of Market Society and Other Recent Essays.* Cambridge, Mass.: Harvard University Press.

Hülvely, I.—Lévai, I. (eds.) (2003): *Regional Politics and Economics of EU Enlargement: East European Regions and Small States in the Changing Architecture of European and Global Integration. European Studies 3.* Budapest: Budapest Forum—Institute for Political Science, HAS, Budapest.

Juma, C. *et al.* (1996): *Open the Social Sciences: Report of the Gulbenkian Commission on the Restructuring of the Social Sciences.* Stanford, Calif.: Stanford University Press.

Kulkinski, A. (ed.) (1997): *European Space—Baltic Space—Polish Space, Part Two, Europe 2010 Series, Vol. 4.* Hannover: Akademie für Raumforschung und Landesplanung (ARL); Warsaw: European Institute for Regional and Local Development (EUROREG).

Laszlo, E. (1977): *Goals for Mankind. Report to the Club of Rome on the New Horizons of Global Community.* New York: E.P. Dutton.

Laszlo, E. (1997): *3rd Millennium. The Challenge and the Vision: The Creative Paths of Human Evolution. Report to the Club of Budapest.* London: Gaia Books.

Lévai, I. (1978): Economic Growth and Foreign Trade in Developing Countries. *Economic Papers*, No. 7. 90-100.

Lévai, I. (1984): Capitalist International Trade: the Contradiction of Equality and Reciprocity. In: Manh-Lan, 1984: 68–78.

Lévai, I. (1986): On the Value and Price of Production of Labour Power. *EADI-Bulletin*, No. 2. 53-64.

Lévai, I. (1988): The Value and Price of Production of Labour Power or the Unequal Exchange Controversy. *Artha Vijnana,* Vol. 30. No. 4. 373-384.

Lévai, I. (1989): Unorganized Employment and Organized Unemployment in India. *Asian Journal of Economics and Social Studies,* Vol. 8. No. 3. 185-210.

Lévai, I. (1996): End of the Long Wave or Bifurcation and Chaos? Seven Points on Global Political Economy. *World Futures*, Vol. 47. 319–323.

Lévai, I. (1998): Calm before the Storm. Some Current Issues of Global Political Economy. *World Futures*, Vol. 51. 321–332. (For an earlier and shorter version of it see: Pending Issues of Global Political Economy. A Research Note. In: Kuklinski 1997: 95-198.)

Lévai, I. (2001): Coexistence of Civilisations and Patterns of Accumulation: the European Sub-Centre and Semi-Periphery. *Central European Political Science Review,* Vol. 2. No. 4. 200-216.

Lévai, I. (2002): Sub-Regional Integration of East European Small States in the European Union: Political Economy of Eastward Enlargement of the EU. *Central European Political Science Review,* Vol. 3. No. 8. 59-73.

Lévai, I. (2003): Political Economy of Eastward Enlargement of the European Union: Small States and Sub-Regions in Europe. In: Hülvely—Lévai 2003: 11-35.

Lévai, I. (2004): Global Governance and Regional Integration. A Research Note on Divergent and Convergent Patterns of Global Social Evolution. In: A. Gergely—Lévai 2004: 3-8.

Lévai, I. (2006): Globalisation: Evolution of the Complex World System. *Central European Political Science Review,* Vol. 7. No. 26. 156-171.

Lévai, I. (2007): Small States in the Complex World System and the European Sub-System. *Foreign Policy Review,* Vol. 4. 104-121.

Luhmann, N. (1989): *Ecological Communication.* Chicago: University of Chicago Press.

Malthus, T.R. (1970 [1826 and 1830]): *An Essay on the Principle of Population; and, A Summary View of the Principle of Population.* Harmondsworth: Penguin Books.

Manh-Lan, N. (ed.) (1984): *Unreal Growth. Critical Studies in Asian Development. Vol. 1.* New Delhi: Hindustan Publishing Corporation Press.

Marx, K. (1977 [1894]): *Capital. A Critique of Political Economy.* Vol. 3. Moscow: Progress Publishers.

Marx, K. (1990 [1867]): *Capital. A Critique of Political Economy.* Vol. 1. Harmondsworth: Penguin Books.

Marx, K. (1990): Appendix: Results of the Immediate Process of Production. In: Marx 1990 (1867): 943-1084.

Meadows, D. H. (1991): *The Global Citizen.* Washington, DC—Covelo, CA: Island Press.

Meadows, D.H.—Meadows, D.L.—Randers, J.—Behrens, W.W. (1972): *The Limits to Growth.* New York: Universe Books.

Meadows, D.—Randers, J.—Meadows, D. (2004): *Limits to Growth. The 30-Year Update.* Vermont: Chelsea Green Publishing Company.

Mill, J.S. (1987 [1848]): *Principles of Political Economy with Some of Their Applications to Social Philosophy.* Fairfield, NJ: August M. Kelley, Publishers.

Modelski, G. (2000): World System Evolution. In: Denemark *et al.* 2000: 24-53.

Prew, P. (2003): The 21st Century World-Ecosystem: Dissipation, Chaos, or Transition? In: Dunaway 2003: 203-219.

Prigogine, I.—Stengers, I. (1979): *La nouvelle alliance: metamorphose de la science.* Paris: Gallimard.

Prigogine, I.—Stengers, I. (1984): *Order out of Chaos. Man's New Dialogue with Nature.* Boulder, CO: Random House.

Robertson, R. (1995): Glocalization: Time-Space and Homogeneity-Heterogeneity. In: Featherstone *et al.* 1995: 25–44.

Rosenau, J.N. (1990): *Turbulence in World Politics. A Theory of Change and Continuity.* New York—London—Toronto—Sydney—Tokyo—Singapore: Harvester Wheatsheaf.

Rosenau, J.N. (1997): *Along the Domestic-Foreign Frontier. Exploring Governance in a Turbulent World.* Cambridge: Cambridge University Press.

Straussfogel, D. (1997): A Systems Perspective on World-Systems Theory. *Journal of Geography*, No. 2. March/April, 119–126.

Urry, J. (2003): *Global Complexity.* Cambridge: Polity Press.

Wallerstein, I. (1974): *The Modern World System. I. Capitalist Agriculture and the Origins of the European World-Economy in the Sixteenth Century.* New York: Academic Press.

Wallerstein, I. (1984): *The Politics of the World-Economy. The States, the Movements, and the Civilizations.* Cambridge: Cambridge University Press; Paris: Editions de la Maison des Sciences de l'Homme.

Wallerstein, I. (1991a): *Geopolitics and Geoculture. Essays on the Changing World-System.* Cambridge—New York: Cambridge University Press; Paris: Editions de la Maison des Sciences de l'Homme.

Wallerstein, I. (1991b): *Unthinking Social Science. The Limits of Nineteenth-Century Paradigms.* Cambridge: Polity Press.

Wallerstein, I. (2000): Globalization or the Age of Transition? A Long-term View of the Trajectory of the World-System. *International Sociology*, Vol. 15. No. 2 249-265.

Renewal of Ecological Ethics: Educational Aspects from a Multicultural Approach to Human-Nature Interactivity

Izaskun Petralanda

"El mundo es una flor que comenzó con Dios, quien lo creó bonito y lo cuidó, pero al tiempo se marchitó, de puro abandono se llenó. Ahora todo el mundo llora, por la flor perdida, que solo Dios cuida." *by Unai Paz*	"The world is a flower that began with God, who created it beautiful and cherished it, but after time it withered, and became filled with neglect. Now the whole world cries, for the flower lost that only God cares for." Translation *by Thomas Heyd*

Introduction

The number of deaths and disabilities directly generated by violence (i.e., suicide, homicide, and war) has alarmingly increased over the last fifty years of the twentieth century (WHO, 2002). Projections made by *The Global Burden of Disease and Injury Project* indicate that war and interpersonal violence will be the main cause of death and disease by the year 2020, with depression and psychiatric illness in second place compromising the functional capacity of tens of millions of people worldwide (Bloom, 1999). These data show that, despite all the scientific and social achievements of the modern era, including the increased average human life expectancy at birth from 46.5 years in 1950 to 65.2 years in 2002, the quality of human life, particularly the ability to relate to others, has deteriorated.

Human violence expresses impairment in the *human ability to relate* either to the *self* or to *other(s),* damaging individuals and societies. Violence towards the environment is called *ecoviolence* (Westra, 2008) and reflects what some ecologists describe as "the cognitive schizophrenia of understanding the world in one manner and yet acting as if it were entirely different" (Pinillos, 2005: 236). Such ecoviolence causes the exhaustion of energy and natural resources, worldwide environmental destruction, and climate change that put the whole bio-geological stability of life on planet Earth in serious peril (Fealy, 2007; Brown, 2007; Motherway, 2007; Blaschke, 2007).

The human ability to relate, because it expresses the perceptions that people have from other people or Nature, is deeply related to human dignity and has a tremendous effect on the respect given to human rights and nature's rights. In turn, dignity and rights are central to an ethical analysis that seeks to ameliorate the current problems facing humankind, such as those described above. Such an ethical analysis requires consideration of the ethos of modern civilization in seeking ways to improve the human ability to relate and to avoid violence. As Diez-Hochleitner and Ikeda indicate:

> [T]he history of humanity shows a rising line that traces the undeniably immense material and cultural achievements made over a long and difficult march through millennia … [I]t also shows how often historical trends around the world have been seriously crippled by the ignorance and selfishness of too many … Our task and renewed vocation is to be carriers of hope for the future … (to) become partners in our commitment to contribute to a reversal of the present most dangerous trends (Diez-Hochleitner & Ikeda, 2008: 126).

In this chapter we present some educational guidelines used in two ethics education programs designed to improve the human ability to relate, both to humans and to nature, through an ontological perspective that promotes assertive relational skills and prosocial moral development.

On the Ontology of the Human Ability to Relate: Cultural Diversity

The human ability to relate expresses itself through different *human recognition patterns,* or procedures by which each person recognizes him/her as *self* and differentiates from *other(s)* as *non-self.* According to Honneth (1997), human recognition patterns affect the emotional, cognitive, and social dimensions of personality, each with different *relational needs* (i.e., affective, moral, or social) that are deeply involved in individual and social development.

The type of recognition pattern that a person uses to satisfy his/her relational needs makes it more or less difficult to establish authentic relationships either to-self or to-other(s). Positive or appreciative recognition patterns generate human relationships that are caring, loving, based on solidarity, and respectful of the rights of others, which has the effect of increasing people's self-confidence, self-respect, and self-esteem, and leads to a positive impact on individual and public health. In contrast, negative or depreciative recognition patterns generate human relationships that are exploitative, disrespectful of others' rights, and which deteriorate human dignity as well as the physical and social integrity of individuals and communities (Petralanda, 2007).

The human ability to relate is rooted in the ontological perspective that each person has about the meaning and value of life. Ontological perspectives that base the value of human life on the assumption that there is no relational condition for humans generate recognition patterns that are negative or depreciative, demanding denial, or elimination of the-other(s) as the only approach to the survival of the-self. This *selfish-self* ontology denies the possibility to establish positive relational styles or to achieve solidarity, as in the existentialist's ontological perspective proposed by Sartre, called here the selfish-self ontology of "being-the-only-person-in-the-world," which posits that:

> [O]riginally, the-other possesses me, it is my enemy, I am in danger in front of him: "hell is the others"… loving expectancy is impossible … The-other as such is never my brother … The-other is "our evil twin," "the being form another species" as Sartre said … and this denial and exclusion make human relations an unsurpassed fight, liquidation and endless competition that extends over all orders of existence … as pure polemics or war, without the possibility of "harmony" or positive bonds, of reciprocal implication between the-self and the-other (Gonzalez, 2001: 221).[1]

Societies based on such a selfish-self ontological perspective generate selfish economic and political systems that deny or destroy individuality (i.e., selfish collectivisms) or reject generosity or empathy to anyone but to-self (i.e., possessive individualism or, at its best, illustrated selfishness). In such societies, very easily, the main purpose of human self-realization becomes not only to possess the-self but also the individual possession of goods and things (Etxeberria, 2005) or the "endless avidity … [so] that people, even when satiated, are avid for material goods without knowing what they want" (Zambrano, 2004: 89). Solidarity is understood as "the unification that two or more people feel when they have in common to be dominated by a third, or when two or more people are united to dominate others. (…) Human relations are conceived as a relation among contraries necessarily exclusive" (González, 2001: 222).

Such societies generate economic and political systems that deeply affect Human-Nature relationships because "the self realization projects come to be possessions competition projects, which imply an exploitative relation with nature, breaking any perspective for solidarity towards it" (Etxeberria, 2005: 125). Individuals and communities increasingly start to consider themselves absolute owners and possessors of all Nature's goods, tangible or intangible, for their own enjoyment and for the sole satisfaction of their own needs. It is a perspective that supports the Baconian metaphor of Nature as "Eros"—an enemy to dominate and subjugate. This metaphor has been particularly catastrophic in the management of rainforest resources over the last thirty years, causing disruption of the stability, complexity, and biodiversity of Amazonian ecosystems through indiscriminate soil erosion and mass and species extinctions which will take thousands of years to correct (Moran, 2000).

In contrast with the selfish-self ontology just described, there are other ontological possibilities that call for authentic relationships as the meaning of human life and of the human relational condition, generating positive recognition patterns and *relational styles* (Maslow, 1972; Rogers, 1981; Lévinas, 1995; Etxeberria, 2005; Diez-Hochleitner, 2008; Petralanda, 1993; Negrete, 2004). This is the case of the *Amazonian Aboriginal People* whose recognition patterns are based on the assumption that all living and non-living beings are connected and interrelated. Therefore, selfish exploitation of others, particularly Nature, is not an acceptable way of living (Eguillor, 1994; Mosonyi, 2004; Negrete, 2004; Petralanda, 2005).

Amazonian Aboriginal People's relational styles involve networks of interconnected knowledge, qualitatively different—i.e., biological, cultural, empirical, sociological, psychological, and spiritual—integrated in very complex yet unified and harmonious systems that support stability in the web of life on Earth. Such constructs are called *Amazonian Aboriginal relational Bioculturomas* (AARB). They value and respect the principles of uncertainty, complexity, precaution, and a responsibility to approach relationships while seeking to avoid the disruption of the delicate, dialogue-based, relational equilibrium that must exist among humans, forest spirits, good and evil. The AARBs resemble in their general structure that of *Systems Biology* whose non-reductionistic approach to the study of life seems to be particularly suitable to correct current environmental problems, including climate change (Oltvai & Barabasi, 2002; Trewavas, 2006).

AARBs are based on thousands of years of meticulous observations, experiments, and trials seeking to improve people's quality of life without disrupting the fragile Amazonian ecosystems, biodiversity, or biocomplexity (Estrella, 1996; Moran, 2000). It has been the invention and systematic use of unique environmental conservation strategies, aimed at managing and repairing Amazonian forest ecosystems, that has kept the rate of extinction relatively low (Moran, 2000; Narbaiza, 2004). AARBs seem to be the reason why (contrary to the wilderness myth and in spite of systematic human interventions in South America's rainforests from the Holocene period to date) it has been possible to conserve the Amazon as the largest area of virgin forest of the world producing 40 percent of the oxygen on Earth (Uhuöttöja, 1983; Moran, 2000; Callicott et al., 2007).

From an ontological perspective, AARBs combine the outer and inner human worlds using appreciative recognition and relational patterns that emphasize being-person-*with*-the-world. It is no longer the-self *or* the-other(s) but the-self *and* the-other(s). This perspective requires one to leave behind the-selfish-self ontology and to embrace an attitude of unarmed hospitality towards the-other(s), similar to what some Western philosophers have described as "the difference that is not indifference ... the endless obligation (to relate) ... not reduced to my responsibility towards the-other, (but) also my inspiration by the-other" (Gabilondo, 2001: 216). The-other(s) is valued not as an object of the-self's affection or interests but as an openness of the-self, a freely chosen rupture of the-self's limits. This non-selfish-self perspective of being-person-with-the-world opens new possibilities to the human ability to relate and might allow us to solve the "deepest problem of ethics" by balancing fairly rational selfishness and rational benevolence (Aranguren, 1997: 284).

The non-selfish-self perspective also recognizes and appreciates diversity, interdependence, and proximity as an essential part of the web of life. Therefore, the enormous diversity of ways of being-person-with-the-world is not reduced to one universal way by actively "identifying experiences of a specific group of subjects as the paradigmatic case of the human like" (Benhabib, 2006: 176). On the contrary, the world is perceived as a community of living beings that interact among themselves, and therefore authentic interactivity is highly valued. It is a perspective that incorporates care and love as essential traits to fully become person-with-the-world. The-self's freedom is therefore not unlimited because, as is expressed in the *Ethical Code of Pemon People,* "[people's] natural freedom [finds] barriers that must be respected in order to live peacefully in a diversified cosmos" (Gutiérrez, 2001: 342).

The respect for Nature of the non-selfish-self perspective indicates the beginning of a kind of human wisdom that respects the deepest sense of human life, dignity, and freedom. The human-centered Baconian metaphor for Nature is rejected in favor of one proposed by Lovelock (1991)—Nature as Gaia, a living planet, a friend to care for and respect. Upon it, human life and actions acquire a new perspective of time because, as theoretical physicist Dyson proposed:

> [T]he destiny of our species is formed by the consideration of survival in six different scales of time. To survive means to compete successfully in six time scales. In the scale of years, the unit is the individual. In the scale of decades, the unit is the family. In the scale of centuries, the unit is the tribe or nation. In the scale of millennia, the unit is culture. In the scale of tens of millennia, the unit is space. In the scale of eons, the unit is the matrix of all the life on our planet. Each human being is the product of his [or her] adaptations on the six time scales. For that reason conflicting loyalties are deep in our nature. [For] the purpose of surviving it has been necessary that we were loyal to ourselves, to our families, our tribes, our cultures, our species and our planet. If our psychological impulses are complicated it is because they are formed by complicated and conflicting exigencies" (Dyson, 1994: 362).

It is the eonic time scale in which the mutability of nature, its living face, becomes visible and it is with this understanding that sustainable human-nature relationships must be sought.

On the Renewal of Ecological Ethics: A Multicultural Approach

The non-selfish-self perspective of being-person-with-the-world, values interactivity and interdependence, and provides new grounds on which to develop harmonious and sustainable interactions between humankind and nature. Such a perspective might contribute to a substantial renovation of ecological and environmental ethics (*ecoethics*), helping life's well-being to blossom on Earth. This is the relational perspective proposed by the ethics of caring and listening (Gilligan, 1994; Benhabib, 2006), the cordial rationality ethics proposed by Cortina (2006), the deep ecological ethics proposed by Leopold, Naess and Sessions (Kwiatkowska and Issa, 1998; Speranza, 2006), and it also represents Amazonian Aboriginal ecoethics (Petralanda et al., 2004).

This perspective contrasts with the remarkable anthropocentrism of ecoethics when based on the-selfish-self or being-the-only-person-in-the-world ontology, which many ecologists and ecoethicists are seeking to overcome because its ontological and referential center to a certain type of "anthropos":

> [T]he human subject is socially defined by modernity … [therefore] the present situation in human societies, and the state of the planet as a whole, forces [us] to review the model of anthropocentrism … ecological ethics [requires] a fundamental change in the perception … of humankind in relation with … other human beings, the set of living beings with whom we share this existence, with the non-living support systems with which life interrelates (Sosa, 1998: 121).

In the case of the Amazonian Aboriginal ecoethics, the emphasis is on non-anthropocentric interactions and interdependence between humankind and nature, upon which survival patterns and lifestyles are based. For example, for the *Uhuöttöja (Piaroa) People*, the Earth is where "the heart of the world" is located—therefore to hunt, to fish, or to tend crops are not only activities of subsistence but also they are activities filled with meaning through which one can relate and communicate with nature. For the *Yanomami People*, life's meaning is intimately connected with nature, the *urihi* (forest) is not only a set of elements formed of soil, plants, and animals that can be useful for human subsistence but also part of human beings (Eguillor, 1984; Finkers, 1991). To the *Pemon People* nature is as important for human beings as soil is for the tree, because life is an integral phenomenon, or stated differently—"life in its total habitat" (Gutiérrez, 1996).

A renewal of ecoethics demands not simply that we align ethical knowledge with the economic and sociological perspectives of environmental problems, but that we achieve a change in human relational patterns towards nature, away from the anthropocentric view of the selfish-self and closer to the non-selfish-self perspective. Such change is a complex process that requires the articulation of different dimensions of reality (i.e., biological, social, spiritual, and cultural), each of them with a large variety of components, values, relations, interdependencies, and rationalities (i.e., logical/empirical, hermeneutic/symbolic, and poetic), aiming to propose intelligent and innovative models to manage and use nature's resources without risking the stability of life on Earth, at present or in the future.

On Education for Ecoethics Renewal: General Guidelines

As indicated above, the complex process of renewing ecoethics—including conceptual and attitudinal changes in the human ability to relate—could be addressed through specific ethics education programs with a broad cultural perspective that emphasizes a non-selfish self-ontological perspective. In the rest of this chapter, we describe the general guidelines of two ethics education programs developed at the Faculty of Science of the Central University of Venezuela and at the Technical School "Madre Mazzarello" based on the considerations previously described above.

One program was initiated in 1996 at the Technical School "Madre Mazzarello" as a three-year technical degree for teenagers and young adults (16 to 20 years old) from different Amazonian populations and cultures (i.e., Yanomami, Yek'wana, Wöhtuja, Pemòn, Jiwi, Baniva, Bare, Yeral, and Criollo). To date, the program has graduated over a hundred technicians, all of them working in Intercultural Primary Health Care Services in multicultural settings of Amazonas or pursuing higher academic degrees in health related careers (Eguillor et al., 2005; Petralanda et al., 2007).

The other program opened in 1999 as a course on Bioethics for Sciences and Technology for university students (19 to 23 years old) at the Faculty of Science of the Central University of Venezuela and it has served as the academic basis for the creation, in 2007, of the Ethics Unit for Sciences and Technology. Over two hundred students of the natural sciences (i.e., physics, chemistry, mathematics, biology, and computational sciences), law, and education have been enrolled in the course, and it is currently offered as an interfaculty course at the University (Petralanda et al., 2005; Petralanda, 2006).

The educational goal of both programs is to develop the moral personality of students with a focus on strengthening their moral sensitivity, moral judgment, and moral activity, through an intimate ontology built and expressed in quotidian ethical and socio-moral dilemmas, using dialogue and relational skills (Puig, 1996; Petralanda, 2007a). The ethical component of both programs is at the core of every subject studied, so that the ethical dimension of science and technology is constructed jointly with scientific knowledge,[2] and from not only the students' external world, but also from the inner world of each student. This approach to ethics education has required, inevitably, that science classrooms become spaces open to dialogue and interactivity from a non-selfish-self perspective, rather than individualistic spaces to accumulate scientific knowledge from a selfish-self perspective.

The major ethical fundamentals of both programs are the Life and the Responsibility Principles (Jonas, 1995, 2000): the virtues of prudence and kindness, and two values called *Environmental Solidarity* and *Cognitive Elegance.* Both programs favor understanding the complexity of interdependence among the different expressions of life (both human and non-human), within the non-selfish-self perspective of being-a-person-careful-and-loving-with-the-world. Both programs promote specific moral options to protect life such as: respect for the self-realization of not only human but also more-than-human life, nonviolence, openness to new models of sustainable (rather than nature-exhaustive) development, preservation of the diversity and complexity of life, and interactive associations as a strategy to promote cooperative autonomy. Figure 1 shows a simplified diagram of the ethical and educational structure of the programs.

From the perspective of the ethics of principles, both programs refer to life as an epiphenomenon of the Life Principle, characterized by the continuous and spontaneous generation of biological and cultural diversity, both of which coevolved over time, shaping humankind and nature. Therefore, life is studied as a *bioculturoma* that includes different types of scientific and cultural constructs[3] (i.e., nanoentities as genetic systems, macro entities as ecosystems and systems of ideas, metaentities as metapopulations, landscape, and culture).

From the perspective of the ethics of consequences, both programs include the Principle of Responsibility (Jonas, 1995) considered as a derivation of human power and commensurable with its magnitude, which must be pursued from a non-selfish-self perspective on different time scales:

> [R]esponsibility with regard to the criteria of causality of the world, according to which the effects of our actions project into the near or far future, until they fade away … [means that] the moment of decision is no longer the one of the individual and [his or her] own short term action, but [rather] the moment of humankind in global and social action (Jonas, 2000: 323).

Virtues, particularly prudence and kindness, are considered human qualities acquired and strengthened as a result of certain attitudes and moral actions that could be developed through education (Puig, 1996; MacIntyre, 2001). Virtues are discussed in a broad multicultural and historical perspective showing the importance that humankind (including scientists) has assigned to them, since ancient times, in order to develop inner strength and character. Virtues are also considered essential to the development of the human ability to relate. Kindness is understood as *benevolence*, that is to say, as loving interest and concern for all expressions of life (Smith, 1997; John Paul II & Bartholomew I, 2002; Diez-Hochleitner & Ikeda, 2008) *prudence* is considered in the sense provided by Bilbao (2006):

> [L]ike the art of the correct development of the continuous "deliberation-judgment-decision," initiated at the intentional moment and culminated at the effective accomplishment moment ... cannot be interpreted like mediocrity, simple precaution without commitment, absence of decision or evasion of responsibilities ... because it is, in fact, the virtue of those who take the initiative, assume the risk, accept the responsibilities and confront the challenges (Bilbao, 2006: 207).

Both programs include *solidarity*—particularly *environmental solidarity*—and *cognitive elegance* as major values, both defined within the non-selfish-self ontological perspective described. Environmental solidarity, a fundamental value in ecoethics, especially within the program designed for science students, seeks to prevent the excessive anthropocentric centering of scientific research. Care is taken to present a sense of human solidarity, which makes possible the interdependent self-realization of humankind as well as the preservation of nature. It requires developing a prudent and kind willingness to be surprised by nature, as well as a trust in the innate human ability to transcend our own everyday realities and to become intelligent interpreters of the unknown language of nature.

In the case of the Intercultural Primary Health Care Services program, environmental solidarity is presented jointly with human solidarity among cultures and within an integral or holistic health perspective. Clarification and strengthening of the interpersonal and intercultural solidarity is a major educational goal of this program because it was evident from the beginning that the students' preconception about the relationships between health care workers and patients was very different depending on their cultural background. Students from Amazonian Aboriginal cultures, regardless of gender, had a relational style based mainly on proximity, love and respect for patients' lifestyles, centered in the patient more than in the health care provider (Figure 2 A and C). Non-indigenous students, especially males, had a more distant relational style, strongly centered in the academic knowledge of the health care provider and with a lower level of respect for patients' lifestyles (Figure 2B). This could be altered, towards a more caring style, as students advanced through the program (Figure 2D).

Regarding the value of *cognitive elegance*, it is defined as the combination of instrumental and ultimate values characteristic of cognitive and attitudinal competences that are needed to build scientific or technological knowledge. Values included within cognitive elegance were selected by students and faculty members of the Science Faculty, from a list of more than 70 characteristics considered valuable for scientists' work (Petralanda, 2007a). Selected values include those

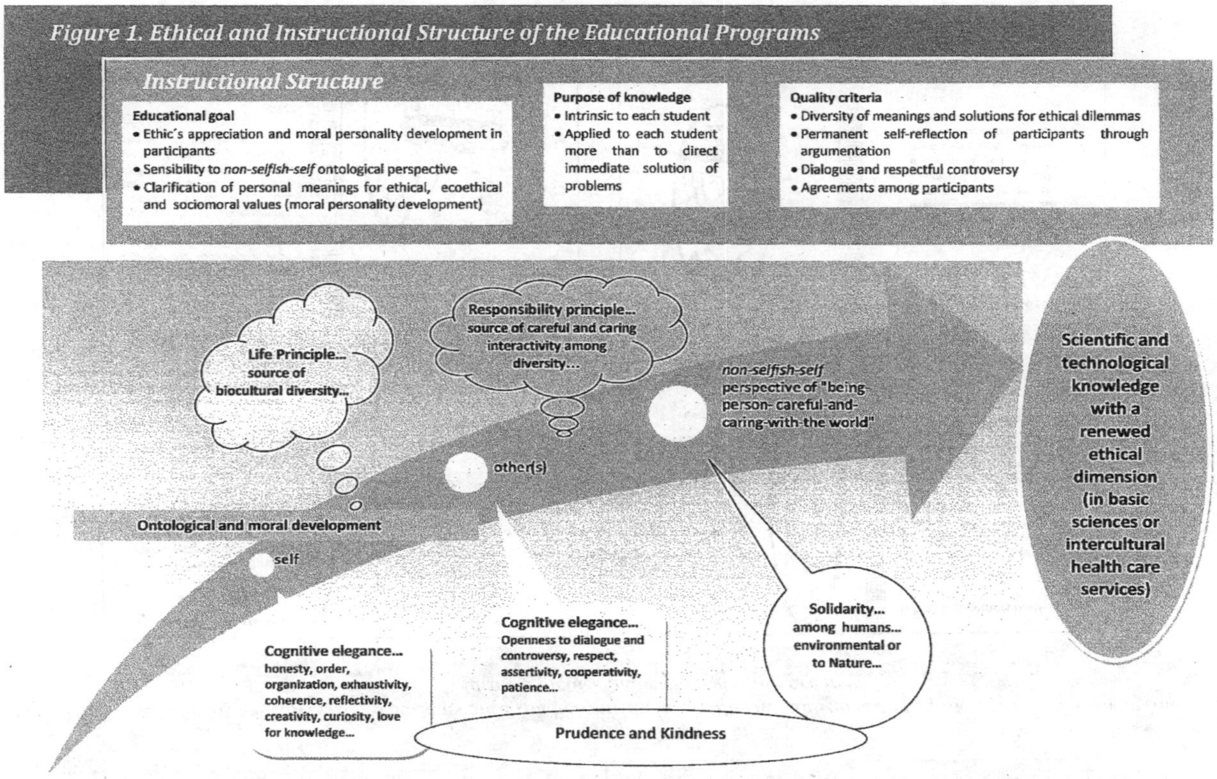

Figure 1. Ethical and Instructional Structure of the Educational Programs

Petralanda, I. (2008) RENEWAL OF ECOLOGICAL ETHICS: Educational aspects from a multicultural approach to Human-Nature interactivity. TODA Conference 2008

Figure 2. Relational styles of Intercultural Public Health Services Program students

Answers to the question

"How would you relate to patients once you become an intercultural public health technician? Please explain and draw a picture indicating who you are in it"

A.Yanomami male student

"When the patient and I first encounter, I would hold to my ethical principles to guide my behavior towards him/her. Specially if he/she is a person that does not belong to my community. I treat the patient, that does not belong to my community, that it is not my family, same as if will by my family, because he/she has the right to know what affects his life, I would trust him/her, I would hold all my patience and care, in other words I would treat him as my family".

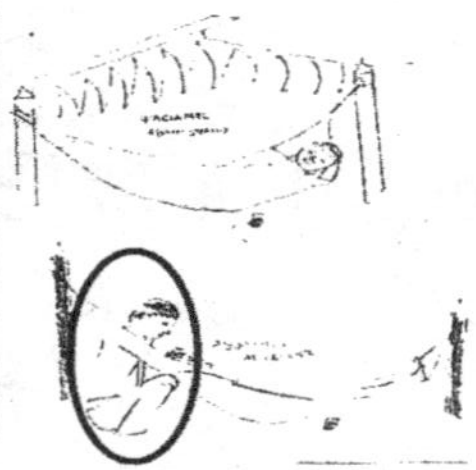

B.Non indigenous male student

"As a future health promoter I would like to attend a sick person in such a way that there could be a respectful, open doctor/patient communication, with responsibility, loyalty and sensitivity. Same way to the relatives that are caring him/her, taking into account that they have the right to know the truth of what is happening to their sick relative"

C. Yek´wana girl student

"I would like to care a patient like to myself in such conditions. First I would let come to me to the patient and the relative accompanying him/her, because he/she not only has a disease but is also suffering and needs somebody care and friendship. I would like to care him/her as in my drawing"

D. Non indigenous female student

"There are not enough words to describe the manner I would like to care a sick person. In summary I would like that my patients feel like in my drawing when I would care them and study their sickness"

Petralanda, I. (2008) RENEWAL OF ECOLOGICAL ETHICS: Educational aspects from a multicultural approach to Human-Nature interactivity. TODA Conference 2008 specific to individual cognitive traits (i.e., order, organization, exhaustivity, coherence, reflectivity, creativity, curiosity, honesty, wisdom (as a love for knowledge)) and those specific to social traits directly linked to the human ability to relate (i.e., patience, cooperativeness, willingness to solve problems, openness to dialogue and controversy, and respect for self and other(s)).

The educational guidelines used to build these programs aim to contribute to the renewal of ethics, including ecoethics, providing ethical support to improve the human ability to relate. They are designed to promote human encounters with other humans and with Nature from a non-selfish-self perspective. They seek to generate joy and cooperative resistance to adversity promoting body, mind and soul well-being. These encounters require generosity and humility, to broaden the-self and to appreciate interdependence among different expressions of life.

Notes

1. English translations of quoted Spanish sources in this manuscript are made by the author.
2. Scientific knowledge is used here as humankind construction of representations searching to know, understand and transform reality (Echeverrìa, 1998).
3. We keep here the double notation to differentiate science as a collective *cultural* construct to which any scientist could contribute regardless of his/her cultural origin from other more particular cultural constructs.

References

Aranguren, J. L. (1997). *Ética.* Madrid: Biblioteca Nueva.

Benhabib, S. (2006). *El Ser y el Otro en la ética contemporánea.* Barcelona: Gedisa.

Bilbao, G.; Fuertes, J., Guibert, J. M. (2006). *Ética para ingenieros.* Bilbao: UNIJES-Desclee de Brouwer.

Blaschke, J. (2007). *La rebelión de Gaia. La verdad sobre el cambio climático.* Bogotá: Robinbook.

Bloom, B. (1999). "The spread of depression: Public health trends." *International Herald Tribune*, Dec 3.

Brown, D. (2007). "Ethical Dimensions of Climate Change." *Proc Environmental Ethics Policy Conference.* Dublin: Royal Dublin Society.

Callicot, J. B.; Rozzi, R.; Delgado, L.; Monticino, M.; Acevedo, M.; Harcombe, P. (2007). "Biocomplexity and conservation of biodiversity hotspots: three case studies from the Americas." *Philosophical Transactions of the Royal Society* 362: 321-333.

Cortina, A. (1996). *Etica mínima.* Madrid: Tecnos.

Diez-Hochleitner R. & Ikeda D. (2008). *A dialogue between East and West. Looking to a Human revolution.* London: I.B. Tauris.

Dyson, F. (1994). *De Eros a Gaia.* Barcelona: Tusquets.

Eguillor, M. I. (1984). *Yopo, shamanes y hekura. Aspectos fenomenológicos del mundo sagrado Yanomami.* Puerto Ayacucho: Ed Salesiana.

Eguillor, M. I., Petralanda, Izaskun, Heredia, Sissy, Moreno, Darìo, Iglesias, Asunciòn (2005). "La educación en ciencia y tecnología en contextos multiculturales: Avances de una investigación didáctica para el desarrollo de contenidos." *Enseñanza de las Ciencias* VII (E).
Echeverrìa, J. (1998). *Filosofìa de la ciencia.* Madrid: Akal.
Etxeberria, X. (2005). *Temas básicos de ética.* Bilbao: Desclee de Brouwer.
Fealy, R. (2007). "Climate Change what does it mean?" *Proc Environmental Ethics Policy Conference.* Dublin: Roy Dublin Soc.
Finkers, J. (1991). "El Yanomami en contacto con su ambiente." *La Iglesia en Amazonas* XII(53), 40-45.
Gabilondo, A. (2001). *La vuelta del otro. Diferencia, identidad, alteridad.* Madrid: Trotta.
Gilligan, C. (1994). *La moral y la teoría: Psicología del desarrollo femenino.* Mexico: FCE.
Gonzalez, J. (2001). *Ética y libertad.* Mexico: UNAM-FCE.
Gutierrez, M. (1996). "Los Pemones." *Etnias Indígenas de Venezuela: Semilla primigenia de nuestra raza.* Caracas: San Pablo.
Honneth, A. (1997). *La lucha por el reconocimiento.* Barcelona: Grijalbo.
Jonas, H. (1995). *El principio responsabilidad: ensayo de una ética para la civilización tecnológica.* Barcelona: Herder.
Jonas, H. (2000). *El Principio Vida. Hacia una biologia filosófica.* Madrid: Trotta.
John Paul II; HH Bartolommeo II (2002). *Common declaration on environmental ethics.*
Kwiatkowska, T.; Issa, J. (1998). *Los caminos de la ética ambiental.* México: Plaza Valdés.
Lévinas, E. (1995). *Totalidad e infinito. Ensayo sobre la exterioridad.* Salamanca: Sígueme.
Lovelock, J. (1991). *Healing Gaia.* New York: Harmony Books.
MacIntyre, A. (2001). *Tras la virtud.* Barcelona: Crìtica.
Maslow, A. (1972). *El hombre autorrealizado.* Barcelona: Kairos.
Morán, E. (2000). *La ecologia humana de los pueblos de la Amazonia.* Mexico: FCE.
Mosonyi, E. (2004). "Reflexiones en torno a la epistemología de la etnociencia." *En la ciencia* 14, 68-78.
Motherway, B. (2007). "Sustainable Energy. The ethical aspects." *Proc Environmental Ethics Policy Conference.* Dublin: Royal Dublin Soc.
Narbaiza, I. (2004) "Consideraciones sobre el desarrollo sostenible en comunidades indígenas del estado Amazonas." *En la ciencia* 14, 124-134.
Negrete, P. (2004). "Problemas epistemológicos de la racionalidad propia de la Cultura andina precolombina." *Mem IV Foro Internacional sobre la espiritualidad de los Pueblos Indios de Amèrica.* México: Universidad de Michoacán.
OMS (2002). *Indicadores de salud mundial.* Geneva: WHO.
Oltvai, Z. & Barabasi, A.-L. (2002). "Life's Complexity Pyramid." *Science* 298: 763-764.
Petralanda, I. (1993). "Reflexiones sobre la muerte del humanismo entre los näpes." *Mem CAICET* III (1-2), 29-38.
Petralanda, I.; Salazar, C.; Ferreira, C.; Gil, E.; González, G.; Rodríguez, A. (2004). "La ética ecológica o ambiental: Situación actual y perspectivas." *Rev Fac Ingeniería* 19(3), 43-50.
Petralanda, I. (2005). "Relaciones entre la diversidad biológica y la diversidad cultural." *Mem I Simp Internacional sobre Pensamiento Andino (Visión estratégica del mundo actual).* Ecuador: UNESCO.

Petralanda, I.; Salazar, C.; Ferreira, C.; Pereyra, E.; Delgado, L.; Gonzàlez, Z.; Hernàndez, J.; Matos, M.; Salazar, M. Winckler, M.; Villarroel, G. (2005). "Ética en la educación científica y tecnológica: Experiencia en la Facultad de Ciencias de la Universidad Central de Venezuela." *Enseñanza de las Ciencias* VII (E).

Petralanda, I. (2006). "Educación ética en la Facultad de Ciencias, UCV." *VI Seminario sobre Problemas Éticos en Venezuela: La dimensión ética de las ciencias y las tecnologías* Caracas: CEI-UCV.

Petralanda, I. (2007). "Ética y bioculturalidad: El reconocimiento del otro(a) en la salud." Pontes Interculturais. Brasil: Nova Harmonia.

Petralanda, I. (2007 a). "Ètica en la educación científica y tecnológica en la Facultad de Ciencias de la UCV: El Principio Vida y la Elegancia cognitiva." Revista Facultad de Humanidades y Educaciòn. Caracas:FHE-UCV.

Petralanda, I.; Eguillor, M. I.; Iglesias, A. (2007a). "Ética y educación intercultural en ciencias y tecnologías: Capacitación de Técnicos Medios en Servicios Interculturales de Salud." Revista Facultad de Humanidades y Educaciòn. Caracas:FHE-UCV.

Pinillos, M. (2005). "La naturaleza histórica de la biodiversidad: Elementos conceptuales de una crisis." *Interciencia* 30(4), 235-242.

Puig, J. (1996). *Construcción de la personalidad moral.* Barcelona: Paidós.

Rogers, C. (1981). *La persona como centro.* Barcelona: Herder.

Smith, P. (1997). *What are they saying about environmental ethics?* New York: Paulist Press.

Sosa, N. (1998). "El qué y el para qué de una ética ecológica." *30 Reflexiones sobre educación ambiental.* Madrid: CENEAM.

Speranza, A. (2006). *Ecología profunda y autorrealización. Introducción a la filosofía ecológica de Arne Naess.* Buenos Aires: Biblos.

Trewavas, A. (2006). "A brief history of Systems Biology." *The Plant Cell* 18: 2420-2430.

Uhuöttöja (1983). "1r Congreso Uhuöttöja." *Bol Antropológico* 7, 42-51.

Westra, L. (2008). *Environmental justice and the rights of indigenous peoples.* New York and London: Earthscan.

Zambrano, M. (2004). *Persona y democracia: la historia sacrificial.* Madrid: Siruela.

The Ethical Implications of Geospatial Technologies and New Perspectives for Community Involvement in the Face of Global Change: The Case of Malaria in Sucre, Venezuela

Laura Delgado Petrocelli

Introduction

Worldwide changes in climate patterns and natural systems, which are often exacerbated by anthropogenic activity, can cause serious impacts on human populations. Modern societies attempt to evaluate the risks to which they are exposed in order to avoid or mitigate their undesirable effects. Two factors define risk: threat and vulnerability. Threat, defined as a latent danger, can be measured as the probability of exceeding the occurrence level of an event, with certain intensity, in a specific site, and during a certain time of exposure (Cardona, 2001, 2003). Vulnerability, on the other hand, is defined as the degree of susceptibility to the physical, economic, political or social damage of a community that could be caused by a destabilizing phenomenon of natural or anthropogenic origin. A population or natural system is considered vulnerable when it displays incapacity to respond to or recover from a certain threat. Vulnerability is expressed as a dynamic system; it can change through time. Natural disasters (earthquakes, tsunamis, hurricanes, mudslides, floods, etc.), including epidemic diseases and anthropogenic threats, can be observed and measured from space. Geospatial technologies play an important role as powerful tools to extract large volumes of information from the spatial context where epidemic diseases develop and allow researchers to integrate this information through geographic information systems (GIS).

At the present time, we see that the emergence, or reemergence, of worldwide infectious diseases and global environmental changes are more closely related than was once believed. Infectious diseases in the world's tropical belt, such as those borne by air (e.g., produced by hantavirus), water (e.g., cholera), and vectors (e.g., dengue fever, malaria, Chagas disease, leishmaniasis, oncocerchiasis, and encephalitis), are particularly active and severe, and have had a strong tendency to reoccur. Malaria is especially prevalent in low-income, rural areas. Malaria emerges and reemerges because of the relaxation of control measures, vector resistance to chemical substances, pathogen resistance to pharmaceuticals, natural environmental deterioration, and human activities. Only recently the disease has been recognized as a complex and dynamic system, therefore a *holistic approach* is needed to counter the various aspects involving malaria's incidence and persistence (Delgado et al., 2004a). This implies a paradigm shift—public health policy-making models must be revised and improved, including adding more criteria, and transdisciplinarian and interdisciplinarian approaches to team work, which can be comprised in a landscape ecology and ecoepidemiological perspective (some call it landscape epidemiology, Kitron, 1998).

In this context, human beings become a landscape element with a measurable impact and traceable history, playing a double role as vulnerable object, and simultaneously as a threatening environmental actor. Human vulnerability usually involves a lack of information concerning surrounding risks, consequences of human activities, and about the implications of research on human habitat that researchers ought to communicate more effectively to individuals and communities. This leads to an interesting and complex situation; how does land use change affect human beings and their communities? If land use changes contribute to the spread of disease by opening new breeding options for disease vectors (malaria in this case), then the human element is a destructive force. Human actions create new grounds for disease persistence, although we may not know what we are doing. Disease resurgence results in community participation, since community members become victims and/or involved community leaders. There might be a public health model to face disease resurgence, but when the model does not work, high-level government officers and regional administrators must participate to activate control programs. When control programs do not work, researchers enter the fray if participation mechanisms are included in the program. There are ethical implications along every step of the ladder. Research findings and potential significance must be transmitted to policy makers, public-health practitioners, regional administrators, community leaders, and finally, to members of the community. After all parties involved have been informed, new proposals from policy makers, partly or completely based on new research findings, shape the new response model (Petralanda & Delgado, 2004). A new ethical paradigm, implying changes in the modes of education, information, socialization, as well as life and environmental values, must be activated at the levels of the individual, the community, and by the authorities, as a new model of public health laws and regulations goes into effect.

Because more than 40 percent of malaria cases in Sucre state (northeastern Venezuela) occur in the Paria peninsula, we studied its southern slope, a 1,078-km^2 area with a large number of wetlands, where recent and increasing land use modification as a result of agricultural activities has been linked to higher incidence of malaria (Delgado, 2005; Delgado & Ramos, 2007). Our research strategy included the use of remote sensing technologies, digital image processing, digital terrain models, global positioning systems, and spatial analysis. Furthermore, the landscape ecology framework allows the integration of human activities, environmental variables, and malaria cases and dynamic behavior. Results derived from the application of these technologies assist in policy design for risk management by health institutions. The overall goal is the design of new improved surveillance and control programs that, if properly implemented, should lead to proper containment of pandemic outbreaks with community participation.

Materials and Methods

Description of Study Area

The study area includes Sucre state, located in northeastern Venezuela between 61°50'44"W, 64°30'00"W, 10°13'10"N, and 10°44'10"N, with a surface of 11,800 km^2 (Figure 1).

Sucre state is located in the tropical belt, generally with high humidity and temperature and a dominant isothermal climate, and high spatial heterogeneity. This implies a differentiated expression for malaria incidence. Epidemiologically, Paria peninsula has become the state's top malaria-infected zone for several years (Figure 2). It was described by Delgado et al. (2000a, 2004a, b, c) as a mountainous range displaying a northward-oriented and a southward-oriented slope. In the southern slope lowlands extend along a gradually declining slope, run-off accumulates in the lowest areas, forming the so-called wetland units. These wetlands contain ideal places for breeding larvae of the malaria disease vector, *Anopheles aquasalis*.

Geospatial Data

A GIS was built from basic cartography at 1:100000 scale and along integrated different thematic layers with epidemiological, socioeconomic, and environmental data. Different basic covers were worked out with MapInfo 7, version 5, software. A digital terrain model (DTM) was constructed at the same scale out of contour lines for the whole Sucre state using the North Wood Geoscience for MapInfo Vertical Mapper, version 3.0 software (Delgado et al., 2003a, 2003c, & 2004c). Covers were UTM coded using the local datum "La Canoa," Zone 20.

Epidemiological Database

Information compiled from the Ministry of Health and Social Development (MSDS) and from the regional Carúpano-Zone XI "Fundasalud Sucre" health management office generated a spatial model of malaria distribution and incidence by locality (population settlements) as a function of number of cases (Delgado et al., 2001).

Socioeconomic Database

Socioeconomic indicators such as population and economically active population (EAP), expressed by type of economic activity (TEA—an indicator of ease of access to employment sources) and a human development capacity index were collected from a study entitled "Typological Characterization and Municipal Classification of Venezuela, 1999" and from the *Atlas of Human Development of Venezuela* (2004) produced in cooperation with the National Institute of Statistics (NIS) and the United Nations Development Program (UNDP). A principal components analysis (PCA) was carried out for the number of cases and malaria indexes as active variables, and illustrative variables describing employment distribution by TEA and other variables related to socioeconomic performance by municipalities from year 1999 to 2000 with the purpose of exploring possible relations between determinant socioeconomic variables and malaria cases (Lebart et al., 1985).

Digital Image Processing

Landsat 153 TM5 December 1990 and Landsat 153 TM7+ September 1999 images were processed out of their BLK format with Er-Mapper, version 5.5, from Earth Resource Mapping, and Erdas Image v 8.6, both

Figure 1
Venezuela (large rectangle contains Sucre State and small one, Paria Peninsula)

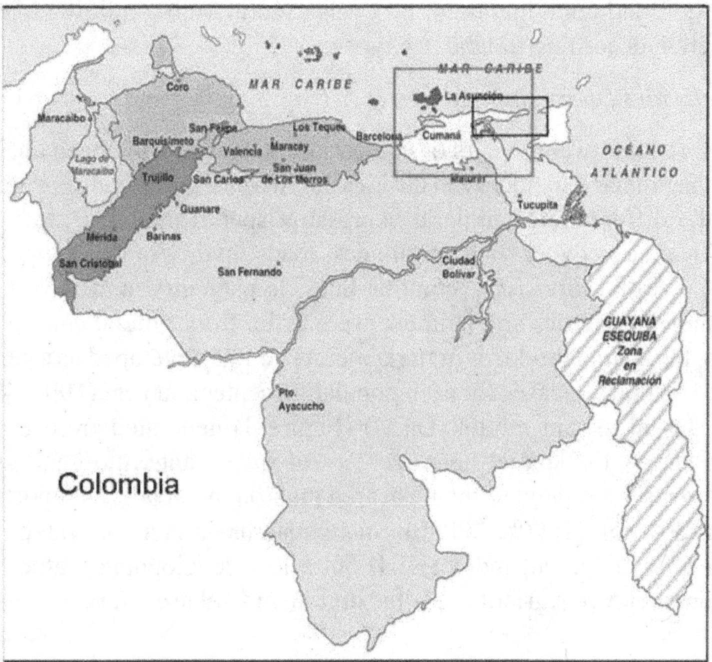

Figure 2
Sucre State and Principal Employment Opportunity by Municipality. Agricultural Activities Lead in the Northeastern Area of Paria Peninsula.

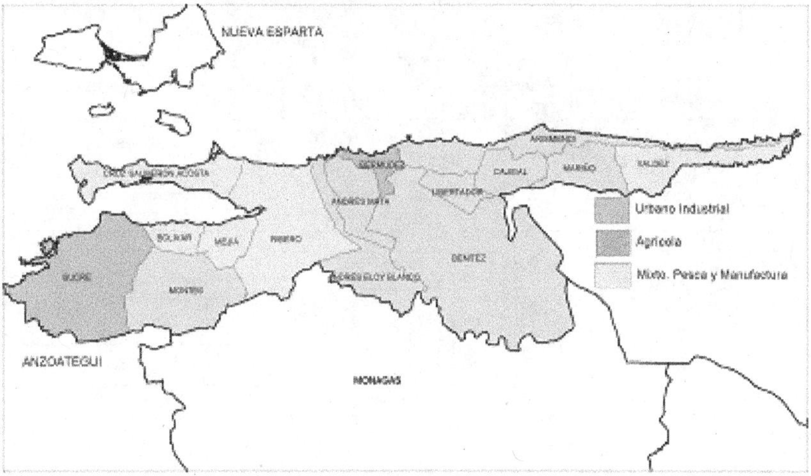

geocoded on terrain control points derived from official digital cartography and then projected on the same reference system used for vector covers. Band combination 4, 5, 3 was used in both to allow rapid unit detection according to land use types.

Results and Discussion

More than twenty years of epidemiological data have been analyzed showing malaria incidence in the area but two years are given as examples for the differentiated malaria expression spatial model. Basic vector and thematic covers such as villages, roads, hydrography, geopolitical administrative divisions, contour lines, hypsometry, and information on human activities and land use were taken from official cartography. Also, the spatial model of malaria incidence was developed as a surface, based on malaria cases for each population center and year (1990, 1999). The digital terrain model (DTM) (Figure 3) generated from contour lines shows the spatial heterogeneity of Sucre state, the gentle-slope low zones are related to the measured malaria incidence, as reported by Delgado et al. (2000b, 2003b). Sucre state has a medium value in the human development index (HDI) but a low development value at the municipal level. Agriculture is the dominant land use, more than half its

Figure 3
Digital Terrain Model

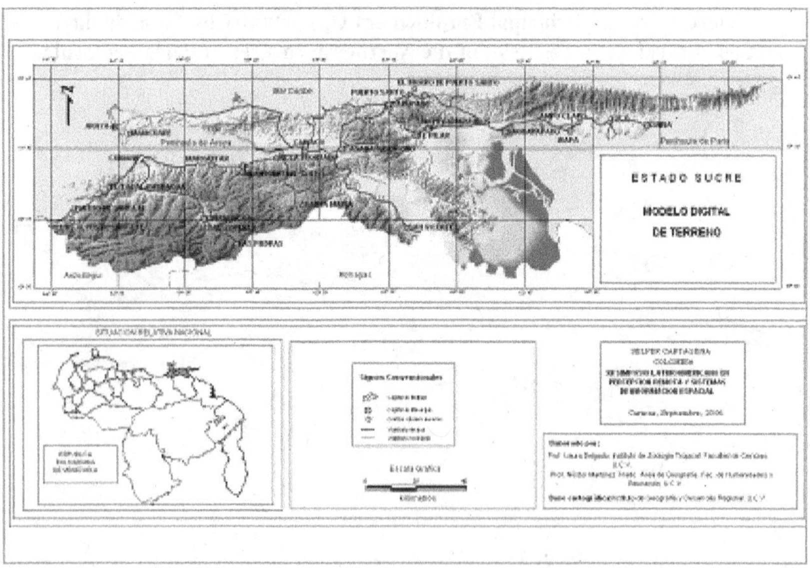

population is affected by critical poverty—basically due to deficiencies in services, a high economic family burden for those employed, and inadequate housing.

Epidemiological characterization based on number of malaria cases and annual parasitic index (API) at state level presents high values for years 1990 and 1999, according to the spatial model of Figure 4.

Figure 4
Spatial Models of Malaria Incidence for Years 1990 and 1999

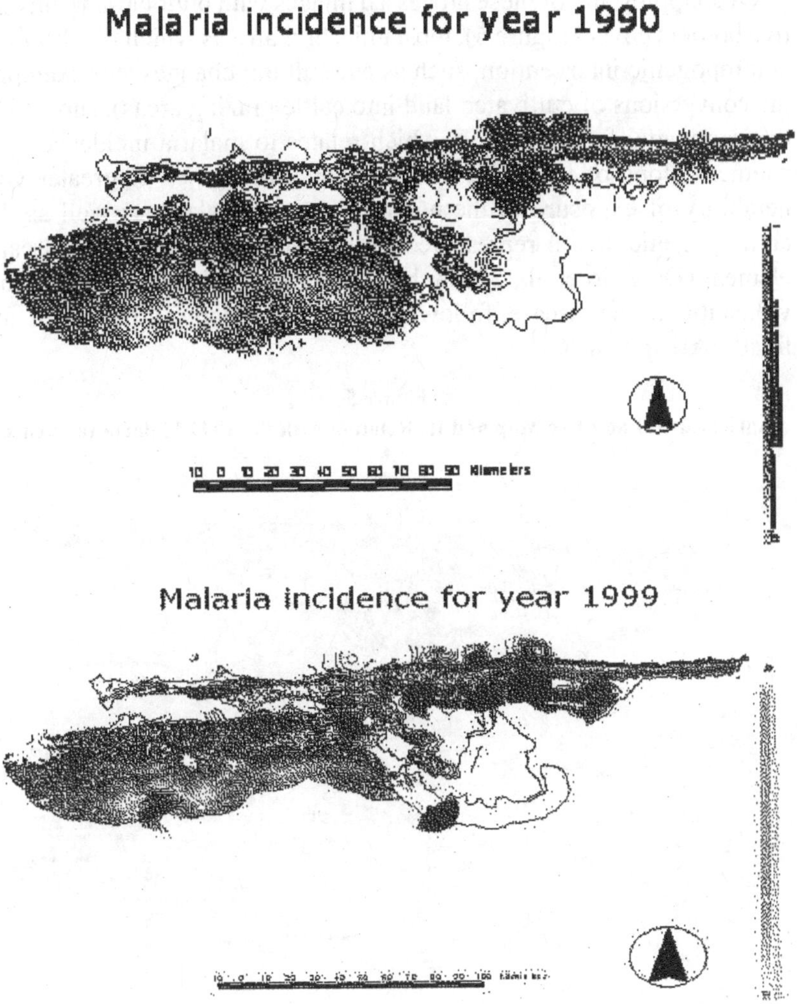

A thematic map was developed from activity and land use structure by municipality and agricultural activity from the National Institute for Statistics (NIS) (Figure 5). The map derived from spatial analysis of the 1999 malaria incidence and agricultural activities land use shows a relation between Cajigal, Mariño, Libertador, and Valdez municipalities and their malaria foci, as reported by Delgado et al. (2004a, c).

Processing of Landsat 153 TM5 and TM7+ images for years 1990 and 1999, respectively, enhanced by band combination 4, 5, 3 allowed land use identification, and detected cover effects due to anthropogenic changes generated each year.

Overlapping any of these processed images with political-administrative borders cover (Figure 6), most units or matrices which display high anthropogenic intervention, such as agricultural changes (for example, the conversions of cultivated land into cattle grazing areas), show high environmental fragmentation, which relates to malaria incidence. The southern slope of the Paria peninsula presents zones of greater vulnerability of exposure to the natural threat of malaria as well as the anthropogenic action represented by human activities as a landscape element (Delgado et al., 2003a, b, c). Cajigal municipality is the most vulnerable to the disease, followed by Mariño, Ribero, Libertador, and finally, Arismendi.

Figure 5
Agricultural Land Use Map and Its Relation with the 1999 Malaria Incidence

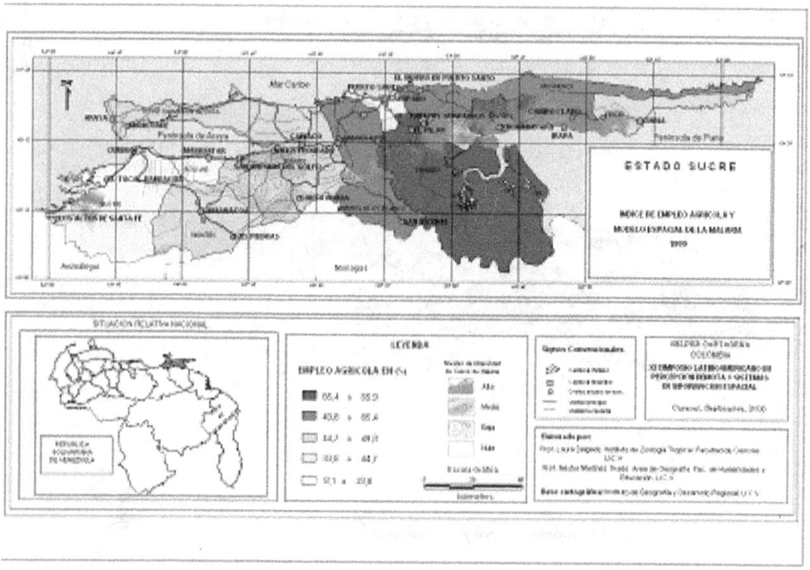

Figure 6
Landsat 153 TM7+ Image (September 1999) Showing Administrative and Political Divisions

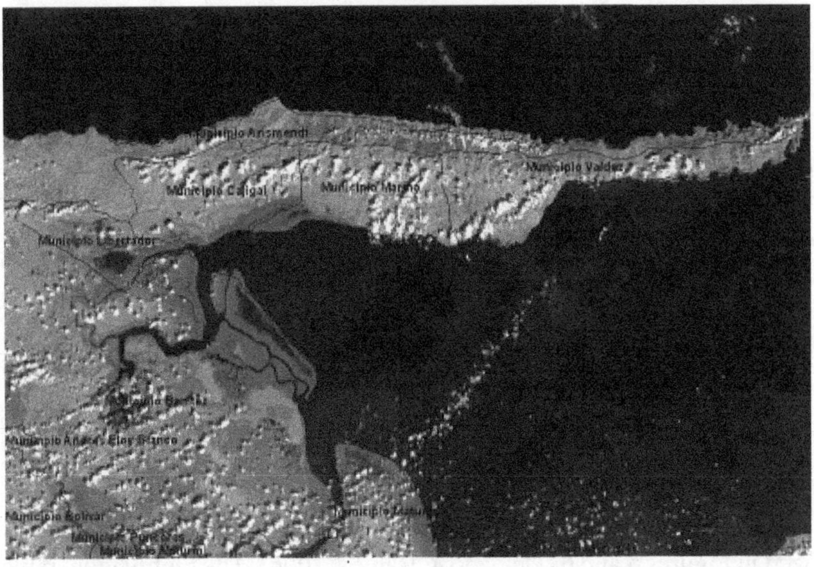

Applying principal components analysis (PCA) to the number of malaria cases and API values for years 1999 and 2000, as active variables, we obtain a first factorial plane (Figure 7).

The first factor explains 80.8 percent of the total inertia and it is possible to interpret this as a size factor that captures malaria incidence and ranks municipalities from more to less malaria. Additionally, the factor can be taken as a vulnerability index, because as defined by Cardona (Cardona, 2001, 2003), it appraises the susceptibility facing a threat and in addition it suggests the difficulty of recovery given the high correlation between API and malaria cases in those consecutive years (Pearson's correlation is r=1 for API and r=0.76 for cases). We can also consider ranking municipalities throughout the first factor according to its vulnerability to malaria. The second factor explains 13.3 percent of total inertia separating municipalities in relation to the absolute and relative importance (by 10,000 inhabitants) of malaria cases. A strong association between the percentage of the population employed in agriculture and API exists and is inversely related to employment in other fields (manufacturing, services, banking, communication, etc.). The spatial relationship between agricultural employment and the zones of greater

Figure 7
Analysis of Principal Components: Distribution of Municipalities in the First Factorial Plane

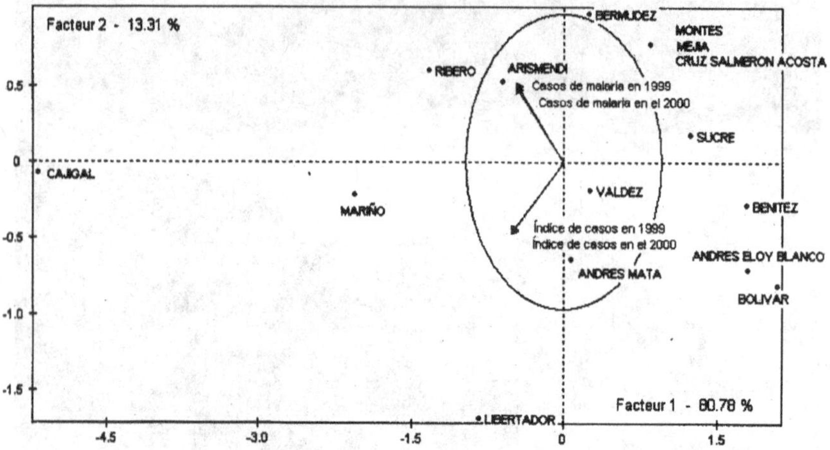

impact by land use change and matrix fragmentation are clearly visualized in Figures 5 and 6 (Delgado & Ramos, 2006). The highly vulnerable municipalities left of the origin in Figure 7 are spatially heterogeneous and in low zones with gentle slope, favoring wetland formation and breeding of the malaria vector, as seen in satellite images (Figure 6). These characteristics increase the vulnerability of the human populations in these municipalities.

This vulnerability is compounded because threatened populations are often ignorant of their risk, lacking information from community leaders, regional authorities, local teachers, researchers, or the media. Through education and socialization of information, communities can process information that could translate into informed program implementation and an improved quality of life. Decision-making and effective action involves many variables: the location of at risk populations, community assistance, treatment type, medicine quality and proper dosage, larvicide treatment type in bodies of water, and natural system impact mitigation measures. Given the complex nature of cooperation and collaboration between all involved parties for developing and implementing an optimal control program, a synergy combining spatial technologies and an approach based on ethical-values is required. Information transmission is vital, both from researchers and from popular culture. Such transmission is sometimes critical for valuable insight in model development. When a community improves through learning and cooperation with researchers and policy makers, whole regions improve, and their lower ecological footprint and impact benefit the planet.

Conclusions

Agricultural use and activities are tied to the presence of malaria in Sucre state. Zones where agricultural activity and anthropogenic intervention are higher correspond with more vulnerability to the impacts of malaria incidence. The first resulting factor from principal component analysis provides a synthetic vulnerability index. Remote sensing and GIS are tools that help in the design of management plans, and for environmental assessment, which may be successful in population vulnerability reduction in malaria zones. Management plans must include an ethical platform that ensures the availability of information to all levels of society, culminating in the adequate education of the most vulnerable subjects. Informed consent from public health program end users provides a minimum of security for unsolicited or inadequate control measures. Activation of ethical values (love for life and environment) through education at the individual and community levels heightens environmental and epidemiological awareness and interconnection between research, policy-making, and communities. Community involvement generates cooperation networks and regional program implementation. Networks facilitate information dissemination and feedback. This chapter has shown how technology can encourage a more systemic approach, and a more democratic and community-based implementation of programs.

Acknowledgments

Special thanks to colleagues Alberto Camardiel, Néstor Martínez, and Santiago Ramos for their valuable support throughout this research. I am indebted to Jorge Zegarra for assistance with the manuscript. Funding was provided by projects Fonacit UCV-2001001850 and CDCH 03-31-3981-2005, second phase.

References

Cardona, D. (2001). *Estimación holística del riesgo sísmico utilizando sistemas dinámicos complejos*. Doctoral Thesis, Universidad Politécnica de Cataluña. pp. 301 http://www.desenredando.org/public/varios/2001/ehrisusd/index.html.

Cardona, D. (2003). *The need for rethinking the concepts of vulnerability and risk from a holistic perspective: A necessary review and criticism for effective risk management*. http://www.desenredando.org/public/articulos/2003/nrcvrfhp/index.html.

Delgado, L., Gamboa, L., & Leon, N. (2000a). "Aspectos geográficos relacionados con un problema de salud pública: La malaria en el estado Sucre." *Revista Terra Nueva Etapa*. Publicación del Instituto de Geografía y Desarrollo Regional, Universidad Central de Venezuela. *16* (25), 80-93.

Delgado, L., Ramos, S., & Gordon, E. (2000b). "Modelo digital del terreno y procesamiento de imágenes digitales: Herramientas para la detección de variables ambientales relacionadas con la malaria en el estado Sucre, Venezuela." IX Simposio Latinoamericano de Percepción Remota, Puerto Iguazú, Misiones, *Memorias SELPER*, pp. 848-856.

Delgado, L., Ramos, S., Rodríguez, R., & Liberal, L. (2001). "Modelo espacial de la probabilidad de riesgo malárico en el edo. Sucre, Venezuela." *Revista de la Facultad de Ingeniería*, Universidad Central de Venezuela, *16* (2), 15-26.

Delgado. L, Ramos, S., Gordon, E., Zoppi, E., Berti, J., & Montiel, E. (2002). "Sensores remotos, modelos digitales de terreno y sig: Herramientas para el control de la malaria, en el estado Sucre, Venezuela." Congreso Internacional de Geodesia y Cartografía Caracas 2002. *Memorias* en formato digital.

Delgado, L., Ramos, S., Martinez, N., & Gamboa, L. (2003a). "La malaria en el estado Sucre: Caso de estudio sobre la relevancia de los métodos geoespaciales en problemas de salud pública." *Acta Científica Estudiantil, 1* (3), 83-95.

Delgado, L., Ramos, S., Martínez, N., & García, P. (2003b). "Ecología de paisajes, sensores remotos y sistemas de información geográfica: Nuevas perspectivas para el manejo de problemas en salud pública, caso particular la malaria en el estado Sucre, Venezuela." *Acta Científica Estudiantil, 1* (4), 128-142.

Delgado, L., Ramos, S., Gordon, E., Zoppi, E., Berti, J., & Montiel, E. (2003c). "Sensores remotos, modelos digitales de terreno y sig: Herramientas para el control de la malaria, en el estado Sucre, Venezuela." *Acta Científica Estudiantil, 1* (4), 143-151.

Delgado, L., & Martínez, N. (2004a). "Relevancia de la variable socioeconómica actividad y empleo agrícola en la dinámica de la malaria. estado Sucre Venezuela." XI Simposio Latinoamericano en Percepción Remota y Sistemas de Información Espacial (SELPER). Santiago de Chile, Chile, *Memorias Selper* en formato digital.

Delgado, L., Ramos, S., Gordon, E., Martínez, N., Zoppi, E., Berti, J., & Montiel, E. (2004b). "Nuevas Perspectivas Para El Manejo De Problemas De Salud Pública: Caso Particular La Malaria." *Revista de la Facultad de Ingeniería*, Universidad Central de Venezuela, *19* (3), 51-61.

Delgado, L., Ramos, S., & Martínez, N. (2004c). "Antropogénesis y persistencia de malaria, estado Sucre-Venezuela." XI Simposio Latinoamericano en Percepción Remota y Sistemas de Información Espacial (SELPER). Santiago de Chile, Chile, *Memorias Selper* en formato digital.

Delgado, L. (2005). "Ecología de Paisajes, Antropogénesis y Malaria". *Revista Journal Selper Edición Especial 20* (1). pp. 24-33.

Delgado, L., & Ramos, S. (2006). "The remote sensing perspective to focus landscape ecology, the anthropogenic action and the malaria disease." The International Trade Event and Conference for eHealth, Telemedicine and Health ICT. Luxembourg, G.D. of Luxembourg. *Proceedings of Med-e-Tel 2006.* Digital. ISSN1819-186X. pp. 332-339.

Kitron, U. (1998). "Landscape ecology and epidemiology of vector-borne diseases: Tools for spatial analysis." *Journal of Medical Entomology, 35* (4), 435-445.

Lebart, L., Morineaux, A., & Fénelon, J. (1985). *Tratamiento estadístico de datos, métodos y programas: Editorial Maracombo, Barcelona, España.* p. 520.

Petralanda, I., & Delgado, L. (2004). "La malaria en Venezuela: Estabilización de focos maláricos y manejo terapéutico de casos de malaria." *Revista de la Facultad de Ingeniería*. Universidad Central de Venezuela. *19* (3), 21-29.

Contributors

David Abram is director of the Alliance for Wild Ethics (AWE), and author of *The Spell of the Sensuous: Perception and Language in a More-than-Human World* (Vintage), for which he received the international Lannan Literary Award for Nonfiction. An accomplished storyteller and sleight-of-hand magician who has lived and traded magic with indigenous sorcerers in Indonesia, Nepal, and the Americas, David lectures and teaches widely on several continents.

Robin Attfield taught philosophy at Cardiff University from 1968 to 2009, and was a full professor there from 1992. He has also taught at Christ Church, Oxford, and in Nigeria and Kenya. He has spoken at several UNESCO conferences, and is currently a professorial research fellow at Cardiff University.

Ellen Gaddy is co-director of NewDay Africa, a not-for-profit organization that supports community-based programs for the orphaned, the ill, and the impoverished in East Africa. Ellen is also a graduate student of psychology and social transformation at Saybrook University. Her areas of focus encompass the psychology of sustainability and sustainable monetary systems.

Marjorie Kelly is author of *The Divine Right of Capital*, and former co-founder and publisher of *Business Ethics* magazine. She is with the Tellus Institute in Boston, where she is co-founder of Corporation 20/20, a multi-stakeholder initiative to envision and advocate for alternative corporate designs.

Dharm P.S. Bhawuk is a professor of management, culture, and community psychology at the University of Hawaii, Manoa. Bhawuk has a Ph.D. in organizational behavior and human resource management from the University of Illinois at Urbana-Champaign. He is the recipient of the 2009 H. Smith Richardson Visiting Fellowship from the Center for Creative Leadership.

Susan Mrazek, M.A. is a doctoral candidate in community and culture psychology at the University of Hawaii at Manoa. She received her bachelor's in psychology from the University of Illinois at Urbana-Champaign.

Vijayan P. Munusamy is a researcher at the Center for Creative Leadership (CCL) – Asia Pacific. Vijayan graduated as a mechanical engineer from the University Technology of Malaysia and with a fellowship from the Asian Development Bank and the East-West Center, he completed his MBA, MPA, and Ph.D. in international management from the University of Hawaii at Manoa.

Imre Lévai, Ph.D., D.Sc., is scientific advisor at the Institute for Political Science of the Hungarian Academy of Sciences; associate professor and director of International Studies MA Programme and head of the Department of European Studies at King Sigismund College, Budapest; and member of the teaching staff of the Doctoral School of the Faculty of Economic Sciences, University of Debrecen, Hungary.

Izaskun Petralanda is professor of ethics, Faculty of Science, Central University of Venezuela. She holds a BA in biology from Central University of Venezuela, an MA in education from Catholic University Andres Bello, and an MA and doctorate in tropical public health, Harvard University.

Laura Delgado Petrocelli is professor and researcher of ecology applied to public health, working with remote sensing assisted Geographic Information Systems (GIS), Laboratory of Information Systems and Environmental Ecology Modeling (Laboratorio de Sistemas de Información y Modelaje Ecológico Ambiental, SIMEA), Institute of Tropical Zoology, Faculty of Science, University of Central Venezuela, Caracas.

Deva Temple is chief sustainability officer for the Mana Makai Group and is co-founder and director of MUSE: Make Us a Sustainable Earth! She worked with the Toda Institute for Global Peace and Policy Research, where she served as research assistant and associate editor of *Peace & Policy* from 2007-2010. She holds a MA in global leadership and sustainable development and a graduate certificate in environmental policy from Hawaii Pacific University, and a bachelor of arts in psychology from Lewis & Clark College.

Olivier Urbain is director of the Toda Institute for Global Peace and Policy Research. He holds a PhD in literature from the University of Southern California (USA) and in peace studies from the University of Bradford (UK). He is the editor of *Music and Conflict Transformation: Harmonies and Dissonances in Geopolitics* (2008) and the author of *Daisaku Ikeda's Philosophy of Peace: Dialogue, Transformation and Global Citizenship* (2010), in which he explores the contributions of humanistic psychology, communicative rationality, and cosmopolitan democracy to the enhancement of a culture of peace.